Universitext

D1270685

Universitext

Editors (North America): J.H. Ewing, F.W. Gehring, and P.R. Halmos

Joseph Rotman

Galois Theory

Springer-Verlag
New York Berlin Heidelberg
London Paris Tokyo Hong Kong

Joseph Rotman
Department of Mathematics
University of Illinois at Urbana-Champaign
Urbana, Illinois 61801
USA

Mathematics Subject Classification (1980): 12-01, 12F10

Library of Congress Cataloging-in-Publication Data
Rotman, Joseph J.
    Galois theory / Joseph Rotman.
      p. cm.—(Universitext)
    Includes bibliographical references.
    1. Galois theory. I. Title.
  QA171.R668   1990
  512′.3—dc20   90-9740

Printed on acid-free paper.

Camera-ready copy prepared using LaTeX.
Printed and bound by R.R. Donnelly & Sons, Harrisonburg, Virginia.
Printed in the United States of America.

9 8 7 6 5 4 3 2 1

ISBN 0-387-97305-2 Springer-Verlag New York Berlin Heidelberg
ISBN 3-540-97305-2 Springer-Verlag Berlin Heidelberg New York

To my teacher Irving Kaplansky

Preface

This little book is designed to teach the basic results of Galois theory—fundamental theorem; insolvability of the quintic; characterization of polynomials solvable by radicals; applications; Galois groups of polynomials of low degree—efficiently and lucidly. It is assumed that the reader has had introductory courses in linear algebra (the idea of the dimension of a vector space over an arbitrary field of scalars should be familiar) and "abstract algebra" (that is, a first course which mentions rings, groups, and homomorphisms). In spite of this, a discussion of commutative rings, starting from the definition, begins the text. This account is written in the spirit of a review of things past, and so, even though it is complete, it may be too rapid for one who has not seen any of it before. The high number of exercises accompanying this material permits a quicker exposition of it. When I teach this course, I usually begin with a leisurely account of group theory, also from the definition, which includes some theorems and examples that are not needed for this text. Here I have decided to relegate needed results of group theory to appendices: a glossary of terms; proofs of theorems. I have chosen this organization of the text to emphasize the fact that polynomials and fields are the natural setting, and that groups are called in to help.

A thorough discussion of field theory would have delayed the journey to Galois's Great Theorem. Therefore, some important topics receive only a passing nod (separability, cyclotomic polynomials, norms, infinite extensions, symmetric functions) and some are snubbed altogether (algebraic closure, transcendence degree, resultants, traces, normal bases, Kummer theory). My belief is that these subjects should be pursued only after the reader has digested the basics.

My favorite expositions of Galois theory are those of E. Artin, Kaplansky, and van der Waerden, and I owe much to them. For the appendix on "old-fashioned Galois theory," I relied on recent accounts, especially [Edwards], [Gaal], [Tignol], and [van der Waerden, 1985], and older books, especially [Dehn] (and [Burnside and Panton], [Dickson], and [Netto]). I thank my colleagues at the University of Illinois, Urbana, who, over the years, have clarified obscurities; I also thank Peter Braunfeld for suggestions that im-

proved Appendix 3 and Peter M. Neumann for his learned comments on
Appendix 4.

I hope that this monograph will make both the learning and the teaching
of Galois theory enjoyable, and that others will be as taken by its beauty
as I am.

Joseph Rotman
Urbana, Illinois, 1990

To the Reader

Regard the exercises as part of the text; read their statements and do attempt to solve them. A star before an exercise indicates that it will be mentioned elsewhere in the text, perhaps in a proof. A result labeled Theorem 5 is the fifth theorem in the text; Theorem A5 is the fifth theorem in Appendix 2 (group theory); Theorem B5 is the fifth theorem in Appendix 3 (ruler-compass constructions).

Contents

Contents

Galois Theory

Galois theory is the interplay between polynomials, fields, and groups. The quadratic formula giving the roots of a quadratic polynomial was essentially known by the Babylonians. By the middle of the sixteenth century, the cubic and quartic formulas were known. Almost three hundred years later, Abel (1824) proved, using ideas of Lagrange and Cauchy, that there exists a quintic polynomial whose roots cannot be given by a formula involving algebraic operations based on the coefficients of the polynomial (actually Ruffini (1799) outlined a proof of the same result, but his proof had gaps and it was not accepted by his contemporaries). In 1829, Abel gave a sufficient condition that a polynomial (of any degree) have such a formula for its roots (this theorem is the reason that commutative groups are called abelian), and Galois (1831) gave a necessary and sufficient condition completely settling the problem. We prove these theorems here.

Rings

The algebraic system encompassing fields and polynomials is a commutative ring. We assume that the reader has, at some time, heard the words *ring* and *homomorphism*; our discussion is, therefore, not leisurely, but it is complete.

Definition. A **commutative ring with** 1 is a set R equipped with two binary operations, addition: $(r, r') \mapsto r+r'$ and multiplication: $(r, r') \mapsto rr'$, such that:

(i) R is an abelian group under addition;

(ii) multiplication is commutative and associative;

(iii) there is an element $1 \in R$ with

$$1r = r \quad \text{for all } r \in R;$$

(iv) the distributive law holds:

$$r(s + t) = rs + rt \quad \text{for all } r, s, t \in R.$$

The **additive group** of R is the abelian group obtained from it by forgetting its multiplication.

We will write *ring* instead of "commutative ring with 1."

Examples

1. $\mathbb{Z}$ (the integers); $\mathbb{Q}$ (the rational numbers); $\mathbb{R}$ (the real numbers); $\mathbb{C}$ (the complex numbers).

2. For a fixed positive integer n, define the ring $\mathbb{Z}_n$ of **integers modulo** n as follows. Its elements are the subsets of $\mathbb{Z}$

$$\overline{a} = \{m \in \mathbb{Z} : m \equiv a \bmod n\}$$
$$= \{m \in \mathbb{Z} : m = a + kn \text{ for some } k \in \mathbb{Z}\},$$

where $a \in \mathbb{Z}$ ($\overline{a}$ is called the **congruence class** of $a \bmod n$). Addition and multiplication are given by

$$\overline{a} + \overline{b} = \overline{a + b} \quad \text{and} \quad \overline{a}\,\overline{b} = \overline{ab},$$

and $\overline{1}$ is "one." It is routine to check that these two binary operations are well defined (that is, they do not depend on the choices of representative in $\overline{a}$ and $\overline{b}$) and that they make $\mathbb{Z}_n$ into a ring.

Recall that $\mathbb{Z}_n$ has exactly n elements: if $a \in \mathbb{Z}$, then there is an integer r with $0 \leq r < n$ such that $\overline{a} = \overline{r}$; moreover, the congruence classes $\overline{r}$ for r in the indicated range are all distinct.

3. If R is a ring, define a **polynomial** $f(x)$ with **coefficients** in R (briefly, a **polynomial over** R) to be a sequence

$$f(x) = (r_0, r_1, \ldots, r_n, 0, 0, \ldots)$$

with $r_i \in R$ for all i and $r_i = 0$ for all $i > n$. If $g(x) = (s_0, s_1, \ldots, s_m, 0, 0, \ldots)$ is another polynomial over R, it follows that $f(x) = g(x)$ if and only if $r_i = s_i$ for all i. Denote the set of all such polynomials by $R[x]$, and define addition and multiplication on $R[x]$ as follows:

$$(r_0, r_1, \ldots, r_i, \ldots) + (s_0, s_1, \ldots, s_i, \ldots) = (r_0 + s_0, r_1 + s_1, \ldots, r_i + s_i, \ldots)$$

and

$$(r_0, r_1, \ldots, r_i, \ldots)(s_0, s_1, \ldots, s_j, \ldots) = (t_0, t_1, \ldots, t_k, \ldots),$$

where $t_0 = r_0 s_0$, $t_1 = r_0 s_1 + r_1 s_0$, and, in general, $t_k = \sum r_i s_j$, the summation being over all i, j with $i + j = k$. Let $(1, 0, 0, \ldots)$ be abbreviated to 1 (there are now two meanings for this symbol). It is routine but tedious to verify that $R[x]$ is a ring, the **polynomial ring over** R.

What is the significance of the letter x in the notation $f(x)$? Let x denote the specific element of $R[x]$:

$$x = (0, 1, 0, 0, \ldots).$$

It is easy to prove that $x^2 = (0, 0, 1, 0, 0, \ldots)$ and, by induction, that x^i is the sequence having 0 everywhere except for 1 in the ith spot. The reader may now prove (thereby recapturing the usual notation) that

$$f(x) = (r_0, r_1, \ldots, r_n, 0, 0, \ldots) = r_0 + r_1 x + \cdots + r_n x^n = \sum r_i x^i$$

($r_0 = r_0 1$ if we identify r_0 with $(r_0, 0, 0, \ldots)$ in $R[x]$). Notice that x is an honest element of a ring and not a variable; its role as a variable, however, is given in Exercise 18.

We remind the reader of the usual vocabulary associated with $f(x) = r_0 + r_1 x + \cdots r_n x^n$. The **leading coefficient** of $f(x)$ is r_n, where n is the largest integer (if any) with $r_n \neq 0$; n is called the **degree** of $f(x)$ and is denoted by ∂f; every polynomial except $0 = (0, 0, \ldots)$ has a degree.

A **monic** polynomial is one whose leading coefficient is 1. The **constant term** of $f(x)$ is r_0; a **constant** (polynomial) is either the zero polynomial 0 or a polynomial of degree 0; **linear**, **quadratic**, **cubic**, **quartic** (or **biquadratic**), and **quintic** polynomials have degrees, respectively, 1, 2, 3, 4, and 5.

Recall from linear algebra that a *linear* homogeneous system over a field with r equations in n unknowns has a nontrivial solution if $r < n$; if $r = n$, one must examine a determinant. If $f(x) = (x - \alpha_1) \ldots (x - \alpha_n) = \sum r_i x^i$, then it is easy to see, by induction on n, that

$$r_{n-1} = -\sum \alpha_i$$

$$r_{n-2} = \sum_{i<j} \alpha_i \alpha_j$$

$$r_{n-3} = -\sum_{i<j<k} \alpha_i \alpha_j \alpha_k$$

$$\vdots$$

$$r_0 = (-1)^n \alpha_1 \ldots \alpha_n.$$

The problem of finding the roots α_i of the polynomial $f(x)$ from its coefficients r_i is thus a question of solving a nonlinear system of n equations in n unknowns; we shall see that this problem is not "solvable by radicals" if $n \geq 5$.

Theorem 1. *Let R be a ring.*

(i) *$0r = 0$ for every $r \in R$;*

(ii) $-r = (-1)r$ for every $r \in R$ (where $-r$ is the additive inverse of r; that is, $-r + r = 0$);

(iii) $(-1)(-r) = r$ for every $r \in R$ (in particular, $(-1)(-1) = 1$).

Proof. (i) The distributive law gives

$$0r = (0 + 0)r = 0r + 0r,$$

and subtracting $0r$ from both sides gives $0r = 0$.

(ii) $0 = 0r = (-1 + 1)r = (-1)r + r$; now add $-r$ to both sides of the equation.

(iii)

$$\begin{aligned}0 = 0(-1) &= (-r + r)(-1) \\ &= (-r)(-1) + r(-1) \\ &= (-r)(-1) - r.\end{aligned}$$

Now add r to both sides. $\square$

If R is a ring in which $1 = 0$ and if $r \in R$, then

$$r = 1r = 0r = 0;$$

hence R consists of exactly one element, namely, 0. We do allow this uninteresting example to be a ring, the **zero ring**. We can now see why "dividing by zero" is forbidden. If $a, b \in R$, then a/b, should it exist, is an element of R such that $b(a/b) = a$. In particular, if $a/0$ exists, then it is an element of R with $0(a/0) = a$. But $0(a/0) = 0$, by Theorem 1(i), and this forces R to be the zero ring.

Two types of ring are especially important: domains and fields.

Definition. A (commutative) ring R is a **domain** (or *integral domain*) if it is not the zero ring and the product of any two nonzero elements in R is itself nonzero.

Note that $\mathbb{Z}_6$ is not a domain because $\overline{2} \neq 0$ and $\overline{3} \neq 0$, but $\overline{2}\,\overline{3} = \overline{6} = 0$. Of course, this example generalizes to $\mathbb{Z}_n$ for any composite n; when n is prime, $\mathbb{Z}_n$ is a domain (if $\overline{a}\,\overline{b} = 0$, then n is a divisor of ab; by Euclid's lemma, which holds when n is prime, n is a divisor of a or b; that is, $\overline{a} = 0$ or $\overline{b} = 0$).

Theorem 2. *A nonzero ring R is a domain if and only if it satisfies the cancellation law: if $ra = rb$ and $r \neq 0$, then $a = b$.*

Proof. Assume R is a domain, $r \neq 0$, and $ra = rb$. Then $r(a - b) = 0$. Since R is a domain, $a - b \neq 0$ is untenable; hence $a - b = 0$ and $a = b$.

Conversely, assume the cancellation law holds. If $r \neq 0$, $a \neq 0$, and $ra = 0$, then $ra = 0 = r0$ implies $a = 0$, a contradiction. $\square$

Exercises

1. Let $f(x), g(x) \in R[x]$.

   (i) Show that the constant term of $f(x)g(x)$ is the product of the constant terms of $f(x)$ and of $g(x)$.

   (ii) If R is a domain, then the leading coefficient of $f(x)g(x)$ is the product of the leading coefficients of $f(x)$ and of $g(x)$.

2. If R is a domain and $f(x)$, $g(x)$ are nonzero polynomials in $R[x]$, then

$$\partial(fg) = \partial f + \partial g.$$

   Conclude that if R is a domain, then $R[x]$ is also a domain.

3. Define the ring of polynomials in two variables over R, denoted by $R[x, y]$, as $A[y]$, where $A = R[x]$. Define polynomials in several variables over R by induction, and show that if R is a domain, then so is $R[x_1, \ldots, x_n]$.

*4. An element $u \in R$ is a **unit** if there exists $v \in R$ with $uv = 1$. Prove that if R is a domain and $f, g \in R$ satisfy

$$f = ug \quad \text{and} \quad g = vf,$$

   where $u, v \in R$, then $uv = 1$ and u, v are units.

   (Notice that 2 is not a unit in $\mathbb{Z}$; of course, $2 \cdot \frac{1}{2} = 1$, but $\frac{1}{2} \notin \mathbb{Z}$. On the other hand, 2 is a unit in $\mathbb{Q}$.)

5. Prove the **division algorithm**: If R is a ring, if $f(x)$, $g(x) \in R[x]$, and if the leading coefficient of $g(x)$ is a unit, then there are polynomials $q(x)$ and $r(x) \in R[x]$ (**quotient** and **remainder**) with

$$f(x) = q(x)g(x) + r(x)$$

   and either $r(x) = 0$ or $\partial r < \partial g$. (See Exercise 13.)

6. Prove that the **binomial theorem** holds in any ring R: if $n \geq 1$, then

$$(a + b)^n = \sum \binom{n}{i} a^i b^{n-i},$$

   where $\binom{n}{i}$ denotes the binomial coefficient $n!/i!(n-i)!$. (Hint: First prove that

$$\binom{n-1}{i-1} + \binom{n-1}{i} = \binom{n}{i}.)$$

*7. If p is a prime, prove that p is a divisor of $\binom{p}{i}$ for all $i \neq 0, p$. (Note that 4 is not a divisor of $\binom{4}{2} = 6$.)

8. If $f(x) \in R[x]$, say, $f(x) = r_0 + r_1 x + \ldots + r_n x^n$, define its **derivative** by

$$f'(x) = r_1 + 2r_2 x + \ldots + nr_n x^{n-1}.$$

Prove that

$$[f(x) + g(x)]' = f'(x) + g'(x)$$

and

$$[f(x)g(x)]' = f(x)g'(x) + f'(x)g(x).$$

Definition. A **field** is a nonzero ring R with $R - \{0\}$ a group under multiplication; that is, every nonzero element of R is a unit.

Examples of fields are $\mathbb{Q}$, $\mathbb{R}$, $\mathbb{C}$, and $\mathbb{Z}_p$. Every field is a domain, for $R - \{0\}$ a group implies it is closed under multiplication; the converse is false, for $\mathbb{Z}$ is a domain that is not a field.

Exercises

9. (i) Define a **subring** of a ring R to be a subset S of R which contains 1 and which is closed under subtraction and multiplication. Show that the intersection of any family of subrings of R is a subring.

   (ii) A **subfield** is a subring that is a field. Show that a subset of a ring is a subfield if it contains 1 and if it is closed under subtraction, multiplication, and inverses.

*10. Show that any intersection of subfields is itself a subfield.

*11. Show that every subring of a field is a domain.

The converse of Exercise 11 is true: given a domain R, one can define its **field of fractions** F; it is a field containing R as a subring and it is constructed from R in exactly the same way as the field $\mathbb{Q}$ is constructed from $\mathbb{Z}$.

In detail, let X denote the set of all ordered pairs $(a, b) \in R \times R$ with $b \neq 0$; define a relation on X of "cross multiplication":

$$(a, b) \sim (c, d) \quad \text{if } ad = bc;$$

this is an equivalence relation (one uses the cancellation law in proving transitivity) and the equivalence class of (a, b) is denoted by a/b. Define addition and multiplication (on the set F of all equivalence classes) by

$$a/b + c/d = (ad + bc)/bd$$

and

$$(a/b)(c/d) = ac/bd$$

(note that $bd \neq 0$ because R is a domain). It is straightforward to check that these operations are well defined (that is, they do not depend on the choices of representative) and that F so equipped is a field. Finally, one can identify $r \in R$ with the "fraction" $r/1$, so that R can be viewed as a subring of F.

*12. (i) Show that $\mathbb{Z}_p[x]$ is an infinite domain containing $\mathbb{Z}_p$ as a subring.

(ii) Show that there exists an infinite field containing $\mathbb{Z}_p$ as a subfield.

*13. If R is a domain, then the quotient and remainder occurring in the division algorithm are unique. (There are rings R, e.g. $\mathbb{Z}_4$, for which the corresponding assertion is false.)

14. Show that $R[x]$ is never a field.

15. Show that $\mathbb{Z}_n$ is a field if and only if n is prime.

Homomorphisms and Ideals

Definition. If R and S are rings, then a function $\psi: R \to S$ is a **ring homomorphism** (or *ring map*) if, for all $r, r', 1 \in R$:

$$\psi(r + r') = \psi(r) + \psi(r');$$
$$\psi(rr') = \psi(r)\psi(r');$$
$$\psi(1) = 1.$$

A ring homomorphism $\psi: R \to S$ is an **isomorphism** if ψ is a bijection;[1] in this case, one says that R and S are **isomorphic** and one writes $R \cong S$.

---

[1] A function $\psi: X \to Y$ is an **injection** if $\psi(x) = \psi(x')$ implies $x = x'$ (one also says ψ is *one-one*); ψ is a **surjection** if, for each $y \in Y$, there exists $x \in X$ with $\psi(x) = y$ (one also says ψ is *onto*); ψ is a **bijection** if it is both an injection and a surjection.

Exercises

16. The relation $R \cong S$ is an equivalence relation on the class of all rings.

17. The *natural map* $\mathbb{Z} \to \mathbb{Z}_n$, defined by $a \mapsto \bar{a}$, is a surjective ring map.

*18. If $a \in R$, define $e_a: R[x] \to R$ by $f(x) = \sum r_i x^i \mapsto \sum r_i a^i$ (denote this element of R by $f(a)$); prove that e_a is a ring map (it is called **evaluation** at a). If $f(a) = 0$, then a is called a **root** of $f(x)$.

(This exercise allows one to regard x as a variable ranging over R; that is, each polynomial $f(x) \in R[x]$ determines a function $R \to R$. But look at the next exercise.)

*19. Give an example of distinct polynomials $f(x), g(x) \in \mathbb{Z}_p[x]$ with $f(a) = g(a)$ for all $a \in \mathbb{Z}_p$.

(Distinct polynomials (not all coefficients are the same) may determine the same function; this is one reason for our defining polynomials in such a formal way. Indeed, if F is any finite field (there are such other than $\mathbb{Z}_p$), there are only finitely many functions $F \to F$ but there are infinitely many polynomials. We shall see after Theorem 11 that this exercise is false if $\mathbb{Z}_p$ is replaced by any infinite field.)

*20. (i) Show that the set of all constants in $R[x]$ is a subring $\tilde{R}$ of $R[x]$ and that $f: R \to \tilde{R}$, defined by $r \mapsto (r, 0, 0, \ldots)$, is an isomorphism.

   (ii) Let R be a domain with field of fractions F. Show that $\{r/1 \in F: r \in R\}$ is a subring of F isomorphic to R.

21. If $a \in R$ is a unit in R and if $\psi: R \to S$ is a ring map, then $\psi(a)$ is a unit in S.

22. If $\sigma: R \to S$ is a ring map, then so is $\sigma^*: R[x] \to S[x]$ defined by

$$\sum r_i x^i \mapsto \sum \sigma(r_i) x^i;$$

if σ is an isomorphism, then so is σ^*.

23. If $\psi: R \to S$ is a ring homomorphism, then its image, denoted by $\operatorname{im} \psi$, is a subring of S.

There is a first isomorphism theorem for rings; let us first introduce the analogue of a normal subgroup.

Definition. An **ideal** in a ring R is a subset I containing 0 such that:

(i) $a, b \in I$ imply $a - b \in I$;

(ii) $a \in I$ and $r \in R$ imply $ra \in I$.

Every ring R contains the ideals R itself and $\{0\}$.

Exercises

24. If $\psi: R \to S$ is a ring map, then its **kernel**,

$$\ker \psi = \{r \in R: \psi(r) = 0\},$$

    is an ideal of R.

25. The function $R[x] \to R$, which associates to each polynomial its constant term, is a ring homomorphism. What is its kernel?

26. If $r_0 \in R$, then $\{rr_0: r \in R\}$ is an ideal in R [called the **principal ideal generated by** r_0; it is denoted by (r_0)].

*27. Let u be a unit in a ring R.

    (i) If an ideal I contains u, then $I = R$.
    (ii) If $r \in R$, then $(ur) = (r)$.
    (iii) If R is a domain and $r, s \in R$, then $(r) = (s)$ if and only if $s = ur$ for some unit u in R.

28. The only units in $\mathbb{Z}$ are 1 and -1.

*29. If F is a field, then the only units in $F[x]$ are the nonzero constants.

*30. A ring R is a field if and only if its only ideals are R and $\{0\}$.

31. The intersection of any family of ideals in R is an ideal in R.

*32. If $a_1, \ldots, a_n$ are elements in a ring R, then the set of all linear combinations,

$$I = \{r_1 a_1 + \cdots + r_n a_n: r_i \in R, i = 1, \ldots, n\},$$

    is an ideal in R; indeed, it is the *smallest* ideal containing $a_1, \ldots, a_n$ (that is, if J is any ideal containing $a_1, \ldots, a_n$, then $I \subset J$).

*33. The set of all $f(x) \in \mathbb{Z}[x]$ having even constant term is an ideal in $\mathbb{Z}[x]$; it consists of all the linear combinations of x and 2.

34. A ring homomorphism $\psi: R \to S$ is an injection if and only if its kernel is (0). Conclude (using Exercise 30) that if R is a field and if $S \neq 0$, then ψ must be an injection and $\operatorname{im} \psi$ is a subfield of S isomorphic to R.

Quotient Rings

Let I be an ideal in R. Forgetting the multiplication for a moment, I is a subgroup of the additive group of R; moreover, R abelian implies that I is a normal subgroup, and so the quotient group R/I exists. The elements of R/I are the cosets $r + I$, where $r \in R$, and addition is given by

$$(r + I) + (r' + I) = (r + r') + I;$$

in particular, the identity element is $0 + I = I$. Recall that $r + I = r' + I$ if and only if $r - r' \in I$. Finally, remember that the **natural map** $\pi: R \to R/I$ is the (group) homomorphism defined by $r \mapsto r + I$.

Theorem 3. *Let I be an ideal in a ring R. Then the abelian group R/I can be equipped with a multiplication which makes it a ring and which makes the natural map $\pi: R \to R/I$ a ring homomorphism.*

Proof. Define multiplication on R/I by

$$(r + I)(r' + I) = rr' + I.$$

To see that this is well defined, suppose that $r + I = s + I$ and that $r' + I = s' + I$; we must show that $rr' + I = ss' + I$; that is, $rr' - ss' \in I$. But

$$rr' - ss' = (rr' - rs') + (rs' - ss') = r(r' - s') + (r - s)s'.$$

Now $r' - s' \in I$ and $r - s \in I$, by hypothesis; hence $r(r' - s') \in I$ and $(r - s)s' \in I$, because I is an ideal. Finally, the sum of two elements of I is again in I, so that $rr' + I = ss' + I$, as desired.

It is routine to see that the abelian group R/I equipped with this multiplication is a ring. The formula $(r + I)(r' + I) = rr' + I$ says that $\pi(r)\pi(r') = \pi(rr')$, where $\pi: r \mapsto r + I$ is the natural map; it follows that π is a ring homomorphism. $\square$

Definition. If I is an ideal in a ring R, then R/I is called the **quotient ring** of R **modulo** I.

Here is an example. Let $R = F[x]$, the polynomial ring over a field F; let I be the (principal) ideal consisting of all the multiples (by polynomials!) of some particular polynomial $p(x)$ of degree n. If $f(x) \in F[x]$, then the division algorithm gives $q(x)$ and $r(x)$ in $F[x]$ with

$$f(x) = q(x)p(x) + r(x),$$

where $r = 0$ or $\partial r < n$. Note that $f(x) + I = r(x) + I$, so we may assume that every coset (except I itself) has a representative of degree $< n$. What is the multiplication in R/I? As usual,

$$(f(x) + I)(g(x) + I) = f(x)g(x) + I;$$

but here we may replace $f(x)g(x)$ by its remainder after dividing by $p(x)$.

In particular, consider $F = \mathbb{R}$ and $p(x) = x^2+1$. In $\mathbb{R}[x]/I$, every element has the form $a + bx + I$, where $a, b \in \mathbb{R}$. Moreover,

$$(a + bx + I)(c + dx + I) = (a + bx)(c + dx) + I$$
$$= ac + (bc + ad)x + bdx^2 + I.$$

The division algorithm applied to $f(x) = \alpha x^2 + \beta x + \gamma$ and $p(x) = x^2 + 1$ gives

$$f(x) = \alpha(x^2 + 1) + [\beta x + \gamma - \alpha];$$

it follows that

$$(a + bx + I)(c + dx + I) = ac - bd + (bc + ad)x + I.$$

Now $\mathbb{R}[x]/I$ is actually a field, for it is easy to exhibit the multiplicative inverse of $a + bx + I$ (where $a \neq 0$ or $b \neq 0$), namely, $c + dx + I$, where $c = a/(a^2 + b^2)$ and $d = -b/(a^2 + b^2)$. We let the reader prove that $\mathbb{R}[x]/I \cong \mathbb{C}$. In particular, the "imaginary" number i with $i^2 = -1$ corresponds to the coset $x + I$.

Exercises

35. Let n be a positive integer and let $I = (n)$ be the principal ideal in $\mathbb{Z}$ generated by n. Show that the quotient ring $\mathbb{Z}/I$ is $\mathbb{Z}_n$, the ring of integers modulo n.

36. If R is a ring, let $I = (x)$ be the principal ideal in $R[x]$ generated by x. Show that $R[x]/I \cong R$.

*37. Prove the **Correspondence Theorem for Rings**. If I is an ideal in a ring R, then there is a bijection from the family of all intermediate ideals J, where $I \subset J \subset R$, to the family of all ideals in R/I, given by

$$J \mapsto \pi(J) = J/I = \{a + I : a \in J\},$$

where $\pi: R \to R/I$ is the natural map. Moreover, if $J \subset J'$ are intermediate ideals, then $\pi(J) \subset \pi(J')$. (Compare with Theorem A9.)

38. Prove the **First Isomorphism Theorem**. If $\psi: R \to S$ is a ring homomorphism with $\ker \psi = I$, then there is an isomorphism $R/I \to \operatorname{im} \psi$ given by $r + I \mapsto \psi(r)$. (Hint: The isomorphism of the First Isomorphism Theorem for groups (Theorem A5) also preserves multiplication.)

*39. Let I be an ideal in a ring R, let J be an ideal in a ring S, and let $\psi: R \to S$ be a ring isomorphism with $\psi(I) = J$. Prove that there is an isomorphism $R/I \cong S/J$.

Polynomial Rings over Fields

Theorem 4. *If F is a field, then every ideal in $F[x]$ is a principal ideal.*

Proof. Let I be an ideal in $F[x]$. If $I = \{0\}$, then $I = (0)$ is principal with generator 0. If $I \neq \{0\}$, choose a polynomial $m(x)$ in I having smallest degree; we claim that $I = (m(x))$.

Clearly, $(m(x)) \subset I$. For the reverse inclusion, take $f(x)$ in I. By the division algorithm, there are polynomials $q(x)$ and $r(x)$ with

$$f(x) = q(x)m(x) + r(x),$$

where either $r(x) = 0$ or $\partial r < \partial m$. Now $r(x) = f(x) - q(x)m(x) \in I$; if $r(x) \neq 0$, then we have contradicted $m(x)$ having the smallest degree of all polynomials in I. Therefore $r(x) = 0$ and $f(x) = q(x)m(x) \in (m(x))$.
□

By Exercise 27(ii), one may choose $m(x)$ to be monic (since F is a field).

Definition. A ring R is called a **principal ideal domain** (PID) if it is a domain in which every ideal is a principal ideal.

The preceding theorem shows that $F[x]$ is a PID when F is a field (of course, the reader knows that $\mathbb{Z}$ is another example of a PID). On the other hand, $\mathbb{Z}[x]$ is not a PID (it is left as an exercise that the ideal I consisting of all polynomials having even constant term is not a principal ideal; see Exercise 33).

Definition. Let R be a ring; if $r, s \in R$, then r **divides** s (or s is a *multiple* of r) if there exists $r' \in R$ with $rr' = s$; one writes $r \mid s$ in this case.

If $I = (r_0)$, then r_0 divides every $s \in I$. Note that $r \mid 0$ for every $r \in R$, but that $0 \mid r$ if and only if $r = 0$; also, $r \mid r$ for every $r \in R$, and r is a unit if and only if $r \mid 1$.

Definition. Let F be a field, and let $f(x), g(x) \in F[x]$. The **greatest common divisor** (gcd) of $f(x)$ and $g(x)$ is a polynomial $d(x) \in F[x]$ such that:

(i) $d(x)$ is a common divisor of $f(x)$ and $g(x)$; that is, $d \mid f$ and $d \mid g$;

(ii) if $c(x)$ is any common divisor of $f(x)$ and $g(x)$, then $c(x) \mid d(x)$;

(iii) $d(x)$ is monic.

One often denotes $d(x)$ by $(f(x), g(x))$. If $(f(x), g(x)) = 1$, then $f(x)$ and $g(x)$ are called **relatively prime**.

Note that the gcd d of f and g, if it exists, is unique. If d' is another gcd, then regard it only as a common divisor and use (ii) to obtain $d' \mid d$; similarly, $d \mid d'$ if one regards d merely as a common divisor. By Exercise 4, $d' = ud$ for some unit $u \in F[x]$; that is, $d' = ud$ for some nonzero constant u (Exercise 27). Since d and d' are both monic, however, $u = 1$ and $d' = d$.

Theorem 5. *Let F be a field and let $f(x), g(x) \in F[x]$ with $g(x) \neq 0$. Then $(f(x), g(x)) = d(x)$ exists and it is a linear combination of $f(x)$ and $g(x)$; that is, there are polynomials $a(x)$ and $b(x)$ with*

$$d(x) = a(x)f(x) + b(x)g(x).$$

Proof. By Exercise 32,

$$I = \{a(x)f(x) + b(x)g(x) : a(x), b(x) \in F[x]\}$$

is an ideal in $F[x]$ containing both $f(x)$ and $g(x)$. Since F is a field, $F[x]$ is a PID; choose a monic polynomial $d(x)$ with $I = (d(x))$, and note that d is a linear combination of f and g. Now d is a common divisor of f and g because $f, g \in (d) = I$. Finally, if c is a common divisor, then $c \mid f$ and $c \mid g$; that is, $f = cc'$ and $g = cc''$. Therefore, $d = af + bg = acc' + bcc'' = c(ac' + bc'')$, and so $c \mid d$. $\square$

Example. If $a \in \mathbb{Z}$, then $\overline{a}$ is a unit in $\mathbb{Z}_n$ if and only if $(a, n) = 1$.

If $\overline{a}$ is a unit, then there is an integer s with $\overline{a}\,\overline{s} = \overline{1}$. Hence $as \equiv 1 \bmod n$; that is, $as - 1 = tn$ for some integer t, and so $as - nt = 1$. It follows that $(a, n) = 1$, for any common divisor of a and n must also divide 1.

Conversely, if $(a, n) = 1$, then there exist integers s and t with $as + nt = 1$; hence $\overline{a}\,\overline{s} = \overline{1}$.

Exercises

40. Let R be the ring of all functions $f \colon \mathbb{R} \to \mathbb{R}$ under pointwise operations: if $f, g \in R$, then $f + g \colon a \mapsto f(a) + g(a)$ and $fg \colon a \mapsto f(a)g(a)$. Show that $f(x) = \max\{x, 0\}$ and $g(x) = \min\{x, 0\}$ have no gcd in R.

*41. Let $f(x) = \prod(x - a_i) \in F[x]$, where F is a field. Show that $f(x)$ has no **repeated roots** ($f(x)$ is not a multiple of $(x - a)^2$ for any $a \in F$) if and only if $(f(x), f'(x)) = 1$, where $f'(x)$ is the derivative of $f(x)$.

Corollary 6 (Euclid's Lemma). *Let F be a field. If $(f(x), g(x)) = 1$ and $f(x)$ divides $g(x)h(x)$, then $f(x)$ divides $h(x)$ in $F[x]$.*

Proof. There are polynomials $a(x)$ and $b(x)$ with $1 = af + bg$. Hence $h = afh + bgh$. But $gh = fk$ for some polynomial k, so that $h = f(ah + bk)$ and f divides h. $\quad\square$

The proof of Euclid's lemma just given is just an adaptation of the usual proof of Euclid's lemma in $\mathbb{Z}$; the same is true for the Euclidean algorithm to be proved next. If one is given explicit polynomials $f(x)$ and $g(x)$, how can one compute their gcd? How can one express the gcd as a linear combination?

Theorem 7 (Euclidean Algorithm). *There are algorithms to compute the gcd and to express it as a linear combination.*

Proof. The idea is just to iterate the division algorithm. Consider the list of equations (we abbreviate $f(x)$ to f, for example):

$$
\begin{array}{ll}
f = q_1 g + r_1 & \partial r_1 < \partial g \\
g = q_2 r_1 + r_2 & \partial r_2 < \partial r_1 \\
r_1 = q_3 r_2 + r_3 & \partial r_3 < \partial r_2 \\
r_2 = q_4 r_3 + r_4 & \partial r_4 < \partial r_3 \\
\quad\vdots & \quad\vdots \\
r_{n-2} = q_n r_{n-1} + r_n & \partial r_n < \partial r_{n-1} \\
r_{n-1} = q_{n+1} r_n + r_{n+1} & \partial r_{n+1} < \partial r_n \\
r_n = q_{n+2} r_{n+1}. &
\end{array}
$$

We claim that $d = r_{n+1}$ is the gcd (after it is made monic). First of all, note that the iteration must stop because the degrees of the remainders strictly decrease (indeed, the number of steps needed is less than ∂g). Second, d is a common divisor, for $d = r_{n+1}$ divides r_n and so the $(n+1)$st equation $r_{n-1} = q_{n+1} r_n + r_{n+1}$ shows that $d \mid r_{n-1}$. Working up the list ultimately gives: $d \mid g$ and $d \mid f$. Third, if c is a common divisor, start at the top of the list and work down: $c \mid f$ and $c \mid g$ imply $c \mid r_1$, and so forth. Therefore, d is the gcd. Finally, one finds a and b by working from the bottom of the list upward. Thus $d = r_{n+1} = r_{n-1} - q_{n+1} r_n$ is a linear combination of r_{n-1} and r_n. Combining this with $r_n = r_{n-2} - q_n r_{n-1}$ gives

$$
\begin{aligned}
d &= r_{n-1} - q_{n+1}(r_{n-2} - q_n r_{n-1}) \\
&= (1 + q_n q_{n+1}) r_{n-1} - q_{n+1} r_{n-2},
\end{aligned}
$$

a linear combination of r_{n-2} and r_{n-1}. This process ends with $d = af + bg$ (note that all q_i and r_i are explicitly known from the division algorithm). $\quad\square$

Corollary 8. *Let $F \subset E$ be fields, and let $f(x), g(x) \in F[x] \subset E[x]$. Then the gcd of f and g computed in $F[x]$ is the same as the gcd of f and g computed in $E[x]$.*

Proof. Regard $f(x)$, $g(x)$ as lying in $E[x]$. The Euclidean algorithm computes their gcd in $E[x]$. But the list of equations (obtained by iterating the division algorithm) has all its terms involving polynomials over F, and hence this list is identical to the list obtained when working with $F[x]$. $\square$

Let us consider $R = F[x]/I$, where F is a field and I is the principal ideal generated by some polynomial $p(x)$. If $(f(x), p(x)) = 1$, then there are polynomials $s(x), t(x) \in F[x]$ with

$$s(x)f(x) + t(x)p(x) = 1;$$

in R this equation becomes

$$s(x)f(x) + I = 1 + I.$$

Thus $f(x) + I$ is a unit in R with inverse $s(x) + I$.

Exercises

42. Find the gcd of $x^5 - 2x^2 + 1$ and $x^3 + 3x^2 - x - 3$ and express it as a linear combination.

43. Let $p(x) = x^3 + x + 1 \in \mathbb{Z}_2[x]$ and let $I = (p(x))$. If $f(x) = r + sx + tx^2 \in \mathbb{Z}_2[x]$ is not in I, find a polynomial $a(x)$ with $p(x) \mid a(x)f(x) - 1$ so that, in $\mathbb{Z}_2[x]/I$,

$$(a(x) + I)(f(x) + I) = 1 + I.$$

Definition. Let F be a field, and let $f(x), g(x) \in F[x]$. The **least common multiple** (lcm) of $f(x)$ and $g(x)$ is a polynomial $m(x) \in F[x]$ such that:

(i) $m(x)$ is a common multiple of $f(x)$ and $g(x)$; that is, $f \mid m$ and $g \mid m$;

(ii) if $c(x)$ is any common multiple of $f(x)$ and $g(x)$, then $m(x) \mid c(x)$;

(iii) $m(x)$ is monic.

Exercise

44. Prove that if $f(x), g(x) \in F[x]$, where F is a field, then their lcm is the monic generator of $(f) \cap (g)$.

Definition. A nonzero polynomial $p(x) \in F[x]$ is **irreducible over** F if $\partial p \geq 1$ and there is no factorization $p(x) = f(x)g(x)$ in $F[x]$ with $\partial f < \partial p$ and $\partial g < \partial p$.

Notice that irreducibility does depend on the field F. Thus, $x^2 + 1$ is irreducible over $\mathbb{R}$, but it factors over $\mathbb{C}$. Linear polynomials (degree 1) are irreducible over any field.

Exercises

45. A polynomial $p(x) \in F[x]$ of degree 2 or 3 is irreducible over F if and only if F contains no root of $p(x)$. (This is false for degree 4: the polynomial $(x^2 + 1)^2$, which factors in $\mathbb{R}[x]$, has no real roots.)

*46. Let $p(x) \in F[x]$ be irreducible. If $g(x) \in F[x]$ is not constant, then either $(p(x), g(x)) = 1$ or $p(x) \mid g(x)$.

47. **(Euclid's Lemma)** If $p(x)$ is irreducible and $p(x)$ divides the product $q_1(x) \cdots q_s(x)$, then $p(x)$ divides $q_j(x)$ for some j.

48. (i) Every nonzero polynomial $f(x)$ in $F[x]$ has a factorization of the form
$$f(x) = a p_1(x) \cdots p_t(x),$$
where a is a nonzero constant and the $p_i(x)$ are (not necessarily distinct) monic irreducible polynomials;

   (ii) the factors and their multiplicities in this factorization are uniquely determined.

(This analogue of the fundamental theorem of arithmetic has the same proof as that theorem: if also $f(x) = b q_1(x) \ldots q_s(x)$, where b is constant and the $q_j(x)$ are monic and irreducible, then uniqueness is proved by Euclid's lemma and induction on $\max\{t, s\}$.)

*49. Let $f(x) = a p_1(x)^{k_1} \cdots p_t(x)^{k_t}$ and $g(x) = b p_1(x)^{n_1} \cdots p_t(x)^{n_t}$, where $k_i \geq 0$, $n_i \geq 0$, a, b are nonzero constants, and the $p_i(x)$ are distinct monic irreducible polynomials (zero exponents allow one to have the same $p_i(x)$ in both factorizations). Prove that
$$\gcd(f, g) = p_1(x)^{m_1} \cdots p_t(x)^{m_t}$$
and
$$\mathrm{lcm}(f, g) = p_1(x)^{M_1} \cdots p_t(x)^{M_t},$$
where $m_i = \min\{k_i, n_i\}$ and $M_i = \max\{k_i, n_i\}$.

Remark. Unique factorization into irreducibles holds in the rings $F[x_1, \ldots, x_n]$, where F is a field, and one can prove existence of gcd's and lcm's there using the formulas in Exercise 49 (these rings are not PID's if $n \geq 2$, and gcd's are not linear combinations).

There is an elementary relation between factoring and roots.

Theorem 9. *Let $f(x) \in F[x]$ and let $a \in F$. Then there is $q(x) \in F[x]$ with*

$$f(x) = q(x)(x - a) + f(a).$$

Proof. Use the division algorithm. Dividing $f(x)$ by $x - a$ gives a quotient and a constant remainder (because $x - a$ has degree 1):

$$f(x) = q(x)(x - a) + r.$$

Evaluating at a gives $f(a) = q(a)(a - a) + r = r$. ☐

Corollary 10. *Let $f(x) \in F[x]$. Then $a \in F$ is a root of $f(x)$ if and only if $x - a$ divides $f(x)$.*

Proof. If a is a root of $f(x)$, then $f(a) = 0$, and the theorem gives $f(x) = q(x)(x - a)$. Conversely, if $f(x) = q(x)(x - a)$, then evaluating at a gives $f(a) = 0$ and a is a root of $f(x)$. ☐

Theorem 11. *If F is a field and $f(x) \in F[x]$ has degree n, then F contains at most n roots of $f(x)$.*

Proof. Suppose that F contains $n+1$ distinct roots of $f(x)$, say, $a_1, \ldots, a_{n+1}$. By the corollary, $f(x) = (x - a_1)g_1(x)$ (for some $g_1(x) \in F[x]$). Now $x - a_2$ divides $f(x)$; by Euclid's lemma, $x - a_2$ divides $g_1(x)$, so that

$$f(x) = (x - a_1)(x - a_2)g_2(x).$$

By induction on n, $f(x) = (x - a_1)(x - a_2) \ldots (x - a_{n+1})g_{n+1}(x)$. This cannot be, for the left side has degree n while the right side has degree greater than n. ☐

The last theorem is false for arbitrary rings R; for example, $x^2 - 1$ has four roots in $\mathbb{Z}_8$.

Recall that every polynomial $f(x) \in F[x]$ determines a function $F \to F$, namely, $a \mapsto f(a)$. In Exercise 19, however, we saw that distinct polynomials in $\mathbb{Z}_p[x]$ may determine the same function. This pathology vanishes when the coefficient field is infinite. Let F be an infinite field and let $f(x) \neq g(x)$ in $F[x]$ satisfy $f(a) = g(a)$ for all $a \in F$. Then $h(x) = f(x) - g(x)$ is not the zero polynomial; hence it has a degree, say, n. But each of the infinitely many elements $a \in F$ is a root of $h(x)$, and this contradicts Theorem 11.

Prime Ideals and Maximal Ideals

Definition. An ideal I in a ring R is called **prime** if $I \neq R$ and $ab \in I$ implies $a \in I$ or $b \in I$.

If $p(x) \in F[x]$ is irreducible, then $I = (p(x))$ is a prime ideal, for $a(x)b(x) \in I$ implies that $p(x)$ divides $a(x)b(x)$; by Euclid's lemma, either $p(x)$ divides $a(x)$ or $p(x)$ divides $b(x)$; thus, $a(x) \in I$ or $b(x) \in I$. Since $I \neq R$, because $\partial p \geq 1$, it follows that I is a prime ideal.

Exercises

*50. If $I = (p(x))$ is a prime ideal in $F[x]$, where F is a field, then $p(x)$ is irreducible.

51. The zero ideal in a ring R is a prime ideal if and only if R is a domain.

52. An ideal $I = (n)$ in $\mathbb{Z}$ is prime if and only if $n = 0$ or n is prime.

*53. The ideal I in $\mathbb{Z}[x]$ consisting of all polynomials having even constant term is a prime ideal.

Theorem 12. *An ideal I in R with $I \neq R$ is a prime ideal if and only if R/I is a domain.*

Proof. Let I be a prime ideal. Suppose that $a + I \neq 0$ and $b + I \neq 0$; that is, neither a nor b lies in I. If $(a + I)(b + I) = ab + I = 0$, then $ab \in I$, contradicting I being prime. The converse is just as easy. $\square$

This theorem gives a swift proof of Exercise 53, for it is easy to see that $\mathbb{Z}[x]/I \cong \mathbb{Z}_2$.

Definition. An ideal I in a ring R is a **maximal ideal** if $I \neq R$ and there is no ideal J with $I \subsetneq J \subsetneq R$.

Theorem 13. *An ideal I with $I \neq R$ in a ring R is a maximal ideal if and only if R/I is a field.*

Proof. The Correspondence Theorem (Exercise 37) shows that I is a maximal ideal if and only if R/I has no ideals other than $\{0\}$ and R/I itself; Exercise 30 shows that this property holds if and only if R/I is a field. $\square$

Corollary 14. *Every maximal ideal is a prime ideal.*

Proof. Every field is a domain. $\square$

Exercises

54. The zero ideal in a ring R is a maximal ideal if and only if R is a field.

55. The ideal I in $\mathbb{Z}[x]$ consisting of all polynomials having even constant term is a maximal ideal.

56. If F is a field, then the kernel of an evaluation map $F[x] \to F$ is a maximal ideal.

The converse of the last corollary is false. For example, the ideal (x) in $\mathbb{Z}[x]$ is prime but not maximal because $\mathbb{Z}[x]/(x) \cong \mathbb{Z}$ is a domain but not a field.

Theorem 15. *If R is a principal ideal domain, then every nonzero prime ideal I is a maximal ideal.*

Proof. Assume there is an ideal $J \neq I$ with $I \subset J \subset R$. Since R is a PID, $I = (a)$ and $J = (b)$ for some $a, b \in R$. Now $a \in J$ implies that $a = rb$ for some $r \in R$, and so $rb \in I$. Since I is prime, either $r \in I$ or $b \in I$. If $b \in I$, then $J \subset I$, a contradiction. If $r \in I$, then $r = sa$ for some $s \in R$, and so $a = rb = sab$; hence $1 = sb$ and $J = (b) = R$, by Exercise 27(i). Therefore, I is maximal. $\square$

Corollary 16. *If F is a field and $p(x) \in F[x]$ is irreducible, then $F[x]/(p(x))$ is a field containing (an isomorphic copy of) F and a root of $p(x)$.*

Proof. Since $p(x)$ is irreducible, the principal ideal $I = (p(x))$ is a nonzero prime ideal; since $F[x]$ is a PID, I is a maximal ideal, and so $E = F[x]/I$ is a field. It is easy to see that $a \mapsto a + I$ is an isomorphism from F to $\{a + I : a \in F\} \subset E$ (one usually identifies F with this subfield of E).

Let $\theta = x + I \in E$; we claim that θ is a root of $p(x)$. Write $p(x) = a_0 + a_1 x + \cdots + a_n x^n$, where $a_i \in F$. Then, in E:

$$
\begin{aligned}
p(\theta) &= (a_0 + I) + (a_1 + I)\theta + \cdots + (a_n + I)\theta^n \\
&= (a_0 + I) + (a_1 + I)(x + I) + \cdots + (a_n + I)(x + I)^n \\
&= (a_0 + I) + (a_1 x + I) + \cdots + (a_n x^n + I) \\
&= a_0 + a_1 x + \cdots + a_n x^n + I \\
&= p(x) + I = I
\end{aligned}
$$

because $I = (p(x))$. But $I = 0 + I$ is the zero element of $F[x]/I$, and hence θ is a root of $p(x)$. $\square$

For example, the polynomial $x^2 + 1 \in \mathbb{R}[x]$ is irreducible, and $\mathbb{R}[x]/(x^2+1)$ is isomorphic to the field of complex numbers $\mathbb{C}$.

Definition. A polynomial $f(x) \in F[x]$ **splits over** F if it is a product of linear factors.

Of course, $f(x)$ splits over F if and only if F contains all the roots of $f(x)$.

Theorem 17 (Kronecker). *Let $f(x) \in F[x]$, where F is a field. There exists a field E containing F over which $f(x)$ splits.*

Proof. The proof is by induction on ∂f. If $\partial f = 1$, then $f(x)$ is linear and we can choose $E = F$. If $\partial f > 1$, write $f(x) = p(x)g(x)$, where $p(x)$ is irreducible. If $p(x)$ is linear, then $f(x)$ splits over any field E which contains F and over which $g(x)$ splits; moreover, such a field E does exist for $g(x)$, by induction. If $\partial p > 1$, then the corollary provides a field B containing F and a root θ of $p(x)$. Hence $p(x) = (x - \theta)h(x)$ in $B[x]$. By induction, there is a field E containing B over which $h(x)g(x)$, hence $f(x)$, splits.   □

We now modify the definition of repeated roots appearing in Exercise 41.

Definition. If $f(x) \in F[x]$ splits over a field E containing F, then $f(x)$ has no **repeated roots** if $f(x)$ is not a multiple of $(x - a)^2$ for any $a \in E$.

Using Exercise 41 and Corollary 8, one sees that $f(x)$ has no repeated roots if and only if $(f(x), f'(x)) = 1$.

Finite Fields

Definition. The **prime field** of a field F is the intersection of all the subfields of F.

By Exercise 10, the prime field is actually a (sub)field (in particular, it is not 0 because every subfield contains 1).

Theorem 18. *If F is a field, then its prime field is isomorphic to either $\mathbb{Q}$ or $\mathbb{Z}_p$ for some prime p.*

Proof. Define $\chi: \mathbb{Z} \to F$ by $n \mapsto n1$ (where 1 is the "one" in F); it is easy to see that χ is a ring map. If $I = \ker \chi$, then $\mathbb{Z}/I$ is a domain (because it is isomorphic to a subring of the field F). Therefore, I is a prime ideal, and hence $I = (0)$ or $I = (p)$ for some prime p. If $I = (0)$, then χ imbeds $\mathbb{Z}$ in F. Since the prime field must contain the multiplicative inverse of any nonzero integer, the prime field is isomorphic to $\mathbb{Q}$ in this case. If $I = (p)$, the first isomorphism theorem gives $\operatorname{im} \chi \cong \mathbb{Z}/(p) = \mathbb{Z}_p$, which is a field; hence $\operatorname{im} \chi$ is the prime field of F.   □

Definition. A field has **characteristic** 0 if its prime field is isomorphic to $\mathbb{Q}$; it has **characteristic** p if its prime field is isomorphic to $\mathbb{Z}_p$.

Exercises

57. Let $f(x), g(x) \in F[x]$. Then $(f, g) \neq 1$ if and only if there is a field E containing both F and a common root of $f(x)$ and $g(x)$.

58. If F has characteristic p, then $pa = 0$ for all $a \in F$.

*59. If F has characteristic p, then $(a \pm b)^p = a^p \pm b^p$ for all $a, b \in F$. (Hint: use Exercise 7.)

60.   (i) If F has characteristic p, then $\sigma: F \to F$, given by $\sigma: a \mapsto a^{p^i}$, is a field map for all $i \geq 0$.

   (ii) If F has characteristic p and $f(x) \in F[x]$, then $(f(x))^p = f(x^p)$.

61. Every finite field has characteristic p for some prime p. Exhibit an infinite field of characteristic p. (Hint: Exercise 12.)

62. If F is a field of characteristic 0 and $p(x) \in F[x]$ is irreducible, then $p(x)$ has no repeated roots. (Hint: Consider $(p(x), p'(x))$.)

The following elementary remark is very useful. If F is a subfield of a field E, then the additive group of E may be viewed as a vector space over F (if $e \in E$ and $a \in F$, define the scalar product ae as the product of the two elements a, e under the given multiplication on E). In particular, a finite field E is a vector space over $\mathbb{Z}_p$ for some prime p. If $\{\alpha_1, \ldots, \alpha_n\}$ is an ordered basis of E, then each $a \in E$ has coordinates $(\lambda_1, \ldots, \lambda_n)$ for λ_i in $\mathbb{Z}_p$; therefore, $|E| = p^n$ for some prime p and some positive integer n.

Theorem 19 (Galois). *For every prime p and every positive integer n, there exists a field having exactly p^n elements.*

Proof. If there were a field K with $|K| = p^n = q$, then $K^{\#} = K - \{0\}$ would be a multiplicative group of order $q - 1$; by Lagrange's theorem (Theorem A3), $a^{q-1} = 1$ for all $a \in K^{\#}$. It follows that every element of K would be a root of the polynomial

$$g(x) = x^q - x.$$

We now begin the construction. By Kronecker's theorem, there is a field E containing $\mathbb{Z}_p$ over which $g(x)$ splits. Define $F = \{\alpha \in E : g(\alpha) = 0\}$; that is, F is the set of all the roots of $g(x)$. Since the derivative $g'(x) = qx - 1 = -1$ (because $q = p^n$ and E has characteristic p), it follows that the gcd $(g, g') = 1$, and so $g(x)$ has no repeated roots; that is, $|F| = q = p^n$. We claim that F is a field, which will complete the proof. If $a, b \in F$, then $a^q = a$ and $b^q = b$. Therefore, $(ab)^q = a^q b^q = ab$, and $ab \in F$. By Exercise 59, $(a - b)^q = a^q - b^q = a - b$, so that $a - b \in F$. Finally, if

$a \neq 0$, then $a^{q-1} = 1$ so that $a^{-1} = a^{q-2} \in F$ (because F is closed under multiplication). $\square$

In Corollary 34 we shall see that any two fields of order p^n are isomorphic.

Exercises

63. Construct a field with four elements by adjoining a suitable root of $x^4 - x$ to $\mathbb{Z}_2$.

*64. Give the addition and multiplication tables of a field having eight elements. (Hint: Factor $x^8 - x$ over $\mathbb{Z}_2$.)

65. Show that a field with four elements is not (isomorphic to) a subfield of a field with eight elements.

Irreducible Polynomials

Our next project is to find some criteria that a polynomial be irreducible; this is usually difficult, and it is unsolved in general.

Exercises

*66. Recall that if $\sigma: R \to S$ is a ring map, then $\sigma^*: R[x] \to S[x]$, defined by

$$\sum r_i x^i \mapsto \sum \sigma(r_i) x^i,$$

is a map of rings. Prove that if R and S are domains and if $\sigma^*(p(x)) \in S[x]$ is irreducible and has the same degree as $p(x)$, then $p(x)$ is irreducible over R. (Note that the degree condition is satisfied if $p(x)$ is monic.)

*67. Let $\sigma: \mathbb{Z} \to \mathbb{Z}_p$ be the natural map. Use the preceding exercise with a suitable choice of prime p to show that $f(x) = x^4 - 10x^2 + 1$ is irreducible in $\mathbb{Z}[x]$. **Remark.** An irreducible polynomial in $\mathbb{Z}[x]$ may factor mod p for some prime p. For example, $f(x) = x^3 + 6x^2 + 5x + 25$ is irreducible in $\mathbb{Z}[x]$ because it becomes $x^3 + 2x + 1 \bmod 3$; this latter (cubic) polynomial is irreducible because it has no roots, the only candidates being $\bar{0}$, $\bar{1}$, $\bar{2}$. But $f(x)$ does factor mod 5, for it becomes $x^3 + x^2 = x^2(x + 1)$.

68. If $\sigma: R \to S$ is a ring isomorphism, then $\sigma^*: R[x] \to S[x]$ is also a ring isomorphism. Conclude that if $p(x) \in R[x]$ is irreducible, then $\sigma^*(p(x)) \in S[x]$ is irreducible.

*69. If $c \in R$, where R is a domain, then the map $f(x) \mapsto f(x + c)$ is an automorphism of the ring $R[x]$. Conclude that $p(x)$ is irreducible if and only if $p(x + c)$ is irreducible.

*70. Let $f(x) = a_0 + a_1 x + \cdots + a_n x^n \in \mathbb{Z}[x]$. If r/s is a rational root of $f(x)$, then $r \mid a_0$ and $s \mid a_n$. Conclude that any rational root of a monic polynomial in $\mathbb{Z}[x]$ must be an integer.

71. Test whether the following polynomials factor in $\mathbb{Z}[x]$:

$$3x^2 - 7x - 5; \qquad 6x^3 - 3x - 18; \qquad x^3 - 7x + 1.$$

72. Prove that if $a_0 + a_1 x + \cdots + a_n x^n \in F[x]$ is irreducible, then so is $a_n + a_{n-1} x + \cdots + a_0 x^n$.

We have seen various ways to determine whether a polynomial $f(x)$ in $\mathbb{Z}[x]$ has a factorization in $\mathbb{Z}[x]$, but we are really interested in its factorization over $\mathbb{Q}[x]$.

Definition. A polynomial $f(x) = a_0 + a_1 x + \cdots + a_n x^n \in \mathbb{Z}[x]$ is called **primitive** if the gcd of its coefficients is 1.

Observe that if $f(x)$ is not primitive, then there exists a prime p which divides each of its coefficients.

Lemma 20 (Gauss). *The product of two primitive polynomials $f(x)$, $g(x)$ is itself primitive.*

Proof. Assume that $f(x)g(x) = (\sum a_i x^i)(\sum b_j x^j) = \sum c_k x^k$ is not primitive, so that there is some prime p dividing each c_k. Let a_i and b_j be the first coefficients of $f(x)$ and $g(x)$, respectively, which are not divisible by p. Then the definition of multiplication of polynomials gives

$$a_i b_j = c_{i+j} - (a_0 b_{i+j} + \cdots + a_{i-1} b_{j+1} + a_{i+1} b_{j-1} + \cdots + a_{i+j} b_0).$$

Since each term on the right is divisible by p, it follows that $a_i b_j$ is divisible by p. But Euclid's lemma in $\mathbb{Z}$ implies that p divides either a_i or b_j, and this is a contradiction. $\square$

Lemma 21. *Every nonzero $f(x) \in \mathbb{Q}[x]$ has a unique factorization*

$$f(x) = c(f)f^*(x),$$

where $c(f) \in \mathbb{Q}$ is positive and $f^*(x) \in \mathbb{Z}[x]$ is primitive.

Remark. The positive rational $c(f)$ is called the **content** of $f(x)$.

Proof. Let $f(x) = (a_0/b_0) + (a_1/b_1)x + \cdots + (a_n/b_n)x^n \in \mathbb{Q}[x]$. Define $B = b_0 \cdots b_n$, so that $f(x) = (1/B)g(x)$ for $g(x) \in \mathbb{Z}[x]$. Now define B' as $\pm$ gcd of the coefficients of $g(x)$ (the sign chosen to make B'/B positive). Then $f(x) = c(f)f^*(x)$, where $c(f) = B'/B$ and $f^*(x) = (B/B')f(x)$, is a desired factorization.

Suppose that $f(x) = dh(x)$ is a second such factorization; then $f^*(x) = rh(x)$, where $r = d/c(f)$ is a positive rational. Write $r = u/v$ in lowest terms; that is, u and v are relatively prime positive integers. Then $vf^*(x) = uh(x)$ is an equation in $\mathbb{Z}[x]$. The coefficients of $uh(x)$ have v as a common divisor; by Euclid's lemma in $\mathbb{Z}$, v divides all the coefficients of $h(x)$. But $h(x)$ is a primitive polynomial, hence $v = 1$. A similar argument shows that $u = 1$. Therefore, $r = d/c(f) = u/v = 1$; hence $d = c(f)$ and $f^*(x) = h(x)$. $\square$

Observe that if $f(x) \in \mathbb{Z}[x]$, then $c(f) \in \mathbb{Z}$ (for it is just the gcd of the coefficients of $f(x)$).

Lemma 22. *If $f(x) \in \mathbb{Q}[x]$ factors as $f(x) = g(x)h(x)$, then*

$$c(f) = c(g)c(h) \quad and \quad f^*(x) = g^*(x)h^*(x).$$

Proof. We have $f(x) = g(x)h(x) = [c(g)g^*(x)][c(h)h^*(x)] = c(g)c(h)g^*(x)h^*(x)$. Since $c(g)c(h)$ is a positive rational, and since the product of two primitive polynomials is primitive, the uniqueness of the factorization in the preceding lemma gives $c(f) = c(g)c(h)$ and $f^*(x) = g^*(x)h^*(x)$. $\square$

Theorem 23. *If $f(x) \in \mathbb{Z}[x]$ factors in $\mathbb{Q}[x]$, then it also factors in $\mathbb{Z}[x]$ (into polynomials of the same degree as over $\mathbb{Q}$). Equivalently, if $f(x) \in \mathbb{Z}[x]$ is irreducible over $\mathbb{Z}$, then it is irreducible over $\mathbb{Q}$.*

Proof. Assume that $f(x) = g(x)h(x)$ in $\mathbb{Q}[x]$. Then $f(x) = c(g)c(h)g^*(x)h^*(x)$ in $\mathbb{Q}[x]$, where g^*, h^* are primitive polynomials in $\mathbb{Z}[x]$. But $c(g)c(h) = c(f) \in \mathbb{Z}$ because $f(x) \in \mathbb{Z}[x]$. Therefore, $f(x) = [c(f)g^*(x)]h^*(x)$ is a factorization in $\mathbb{Z}[x]$. $\square$

(The proof of this last theorem can be adapted to a more general situation: replace $\mathbb{Z}$ by a "unique factorization domain" and replace $\mathbb{Q}$ by its field of fractions. This is the main ingredient of the proof that if R is a unique factorization domain, then so is $R[x]$; it follows that if F is a field, then $F[x_1, \ldots, x_n]$ is a unique factorization domain.)

Theorem 24 (Eisenstein Criterion). *Let $f(x) = a_0 + a_1x + \cdots + a_nx^n \in \mathbb{Z}[x]$. If there is a prime p dividing a_i for all $i < n$, but with p not dividing a_n and p^2 not dividing a_0, then $f(x)$ is irreducible over $\mathbb{Q}$.*

Proof. Let $f(x) = (b_0 + b_1 x + \cdots + b_m x^m)(c_0 + c_1 x + \cdots + c_k x^k)$; by Theorem 23, we may assume that both factors lie in $\mathbb{Z}[x]$. Now $p \mid a_0 = b_0 c_0$ so that, by Euclid's lemma in $\mathbb{Z}$, $p \mid b_0$ or $p \mid c_0$; since p^2 does not divide a_0, only one of them is divisible by p, say, $p \mid c_0$ but p does not divide b_0. The leading coefficient $a_n = b_m c_k$ is not divisible by p, so that p does not divide c_k (or b_m). Let c_r be the first coefficient not divisible by p (so p does divide $c_0, \ldots, c_{r-1}$). If $r < n$, then $p \mid a_r$, and $b_0 c_r = a_r - (b_1 c_{r-1} + \cdots + b_r c_0)$ is divisible by p; hence $p \mid b_0 c_r$, contradicting Euclid's lemma (because p divides neither factor). It follows that $r = n$, hence $k = 0$, and the second factor is constant. Therefore, $f(x)$ is irreducible. $\square$

To illustrate the Eisenstein criterion, $x^5 - 4x + 2$ is irreducible over $\mathbb{Q}$. (This polynomial does not surrender easily to our first criterion.)

Definition. If p is a prime, then the pth **cyclotomic polynomial** is

$$\Phi_p(x) = (x^p - 1)/(x - 1) = x^{p-1} + x^{p-2} + \cdots + x + 1.$$

Corollary 25. *The pth cyclotomic polynomial is irreducible over $\mathbb{Q}$ for every prime p.*

Proof. Recall Exercise 69: a polynomial $f(x)$ is irreducible if and only if $f(x + c)$ is irreducible, where c is a constant. In particular, $\Phi_p(x) = (x^p - 1)/(x - 1)$ is irreducible if and only if $\Phi_p(x + 1) = ((x + 1)^p - 1)/x$ is irreducible. The latter polynomial is $x^{p-1} + px^{p-2} + \binom{p}{2} x^{p-3} + \cdots + p$, where $\binom{p}{i}$ is the binomial coefficient. Since p is prime, Exercise 7 shows that Eisenstein's criterion applies; we conclude that $\Phi_p(x)$ is irreducible. $\square$

Corollary 26. *If $a \neq \pm 1$ is a square-free integer, then $x^n - a$ is irreducible over $\mathbb{Q}$ for every $n \geq 2$.*

Proof. Since $a \neq \pm 1$, there is some prime p dividing a, and Eisenstein's criterion applies with this prime. $\square$

This last corollary shows that there are irreducible polynomials over $\mathbb{Q}$ of arbitrary degree n.

Classical Formulas

Let us now derive the classical formulas for the roots of quadratics, cubics, and quartics. Consider the quadratic equation

$$X^2 + bX + c = 0.$$

Replacing X by $x = X - b/2$, this equation becomes

$$x^2 + c - b^2/4 = 0;$$

it follows that $x = \pm\frac{1}{2}\sqrt{b^2 - 4c}$. Of course, one obtains the usual formula by replacing x by $X + b/2$.

A cubic equation $X^3 + aX^2 + bX + c = 0$ becomes, after replacing X by $x = X - a/3$,

$$x^3 + qx + r = 0;$$

as above, a formula for the roots of this equation will give a formula for the roots of the original one. The coming formula is due to Scipio del Ferro (ca. 1515); an equivalent formula was discovered by Tartaglia about the same time; it appeared in print for the first time in the book of Cardan (1545). Set $x = y + z$. Then

$$x^3 = (y + z)^3 = y^3 + z^3 + 3(y^2 z + yz^2) = y^3 + z^3 + 3xyz.$$

Therefore,

(1) $$y^3 + z^3 + (3yz + q)x + r = 0.$$

So far we have imposed only one constraint on y and z, namely, $x = y + z$. Now impose a second constraint:

$$yz = -q/3,$$

so that, in Eq. (1), the linear term in x vanishes. We have

$$y^3 + z^3 = -r$$

and

$$y^3 z^3 = -q^3/27.$$

These two equations can be solved for y^3 and z^3. In detail,

$$y^3 - q^3/27y^3 = -r,$$

and hence

$$y^6 + ry^3 - q^3/27 = 0$$

and

$$z^6 + rz^3 - q^3/27 = 0.$$

The quadratic formula gives

$$y^3 = \frac{1}{2}\left(-r + \sqrt{r^2 + 4q^3/27}\right)$$

and

$$z^3 = \frac{1}{2}\left(-r - \sqrt{r^2 + 4q^3/27}\right).$$

Since $y^3 z^3 = -q^3/27$, we may choose y and z so that $yz = -q/3$. If $\omega = e^{2\pi i/3}$ is a cube root of unity, there are now six cube roots available:

$$y, \quad \omega y, \quad \omega^2 y, \quad z, \quad \omega z, \quad \omega^2 z;$$

these may be paired to give product $-q/3$:

$$-q/3 = yz = (\omega y)(\omega^2 z) = (\omega^2 y)(\omega z).$$

We conclude that the roots of the cubic polynomial are:

$$y + z; \quad \omega y + \omega^2 z; \quad \omega^2 y + \omega z,$$

where

$$y = \left[\frac{1}{2} \left(-r + \sqrt{r^2 + 4q^3/27} \right) \right]^{1/3}$$

and

$$z = \left[\frac{1}{2} \left(-r - \sqrt{r^2 + 4q^3/27} \right) \right]^{1/3};$$

this is the **cubic formula**.

The physicist R. Feynman speculates, in a popular book,[2] that the cubic formula was important in the development of modern science, for it is a result unknown to the ancients. Remember that the year 1515 is virtually simultaneous with Martin Luther and the beginning of the Reformation and the Renaissance.

The quartic formula was found by Luigi Ferrari (ca. 1545), but we present the method of Descartes (1637). Consider the quartic polynomial

$$X^4 + aX^3 + bX^2 + cX + d;$$

replacing X by $x = X - a/4$, we obtain a polynomial of the form

$$f(x) = x^4 + qx^2 + rx + s.$$

Write

$$x^4 + qx^2 + rx + s = (x^2 + kx + \ell)(x^2 - kx + m),$$

where k, ℓ, and m are to be determined (the linear term in the second factor is $-k$ because the quartic has no cubic term). If k, ℓ, and m are known, then the problem is solved by applying the quadratic formula. Expanding the right side and equating coefficients of like terms gives:

$$\ell + m - k^2 = q;$$
$$k(m - \ell) = r;$$
$$\ell m = s.$$

---

[2] R.P. Feynman, "What do you care what other people think?" Further adventures of a curious character, Bantam, 1988, page 95.

The first two equations yield:

$$2m = k^2 + q + r/k;$$
$$2\ell = k^2 + q - r/k.$$

Substituting into the third equation gives

$$k^6 + 2qk^4 + (q^2 - 4s)k^2 - r^2 = 0.$$

This is a cubic in k^2 (essentially the "resolvent cubic" we will meet later), and one can thus solve for k^2 using the cubic formula. It is now easy to determine k, ℓ, and m, and hence to determine the roots of $f(x)$.

It is easy to see why our ancestors were tempted to find a similar formula for a quintic; surely it, too, would yield to ingenuity.

Splitting Fields

We have already observed that if F is a subfield of E, then E may be viewed as a vector space over F.

Definition. If F is a subfield of a field E, one says that E is an **extension field** of F, and one writes "E/F is an extension field." The dimension of E viewed as a vector space over F is called the **degree** of E over F and it is denoted by $[E\colon F]$. One says that E/F is a **finite extension** if $[E\colon F]$ is finite.

Note that the term "extension" inverts one's viewpoint. Instead of focusing on subfields F of E, we focus on larger fields E containing F.

Theorem 27. *Let $p(x) \in F[x]$ be an irreducible polynomial of degree d. Then $E = F[x]/(p(x))$ is an extension field of F of degree d.*

Proof. Denote $(p(x))$ by I, and denote $x + I$ in E by α; it suffices to prove that $\{1, \alpha, \alpha^2, \ldots, \alpha^{d-1}\}$ is a basis of E over F. If, for $0 \leq i \leq d-1$, there are $a_i \in F$ with $\sum a_i \alpha^i = 0$, then α is a root of $f(x) = \sum a_i x^i$, a polynomial of degree $< d$, contradicting $p(x)$ being a polynomial of least degree having α as a root. Hence $\{1, \alpha, \alpha^2, \ldots, \alpha^{d-1}\}$ is independent. Every element of E has the form $f(x) + I$; the division algorithm gives $q(x)$ and $r(x)$ with $f(x) = q(x)p(x) + r(x)$, where $\partial r < \partial p = d$, and $f(x) + I = r(x) + I$. Hence $\{1, \alpha, \alpha^2, \ldots, \alpha^{d-1}\}$ spans E, and so it is a basis. $\square$

Definition. Let E/F be an extension field, and let $\alpha_1, \ldots, \alpha_n \in E$. Then $F(\alpha_1, \ldots, \alpha_n)$ is the smallest subfield of E containing F and $\alpha_1, \ldots, \alpha_n$; it is called the field obtained by **adjoining** $\alpha_1, \ldots, \alpha_n$ to F. An extension E/F is a **simple extension** if there exists $\alpha \in F$ with

$$E = F(\alpha) = \{f(\alpha)/g(\alpha)\colon f(x), g(x) \in F[x] \text{ and } g(\alpha) \neq 0\}.$$

Definition. Let E/F be an extension field, and let $\alpha \in E$. Then α is **algebraic over** F if α is a root of some monic polynomial $\in F[x]$; otherwise α is **transcendental over** F. An extension E/F is called **algebraic** if every element of E is algebraic over F.

When one says that π or e is transcendental, one usually means that they are transcendental over $\mathbb{Q}$. If F is a field, let $F(x)$ denote the field of all **rational functions** over F; it is the field of fractions of $F[x]$, and its elements are all $f(x)/g(x)$, where $f(x), g(x) \in F[x]$. In this case, x is transcendental over F.

Exercises

*73. If σ is an isomorphism of $F(\alpha_1, \ldots, \alpha_n)$ with itself such that $\sigma(\alpha_i) = \alpha_i$, for $i = 1, \ldots, n$, and $\sigma(a) = a$ for all $a \in F$, then σ is the identity. If $\sigma, \tau : F(\alpha_1, \ldots, \alpha_n) \to E$ fix F pointwise and $\sigma(\alpha_i) = \tau(\alpha_i)$ for all i, then $\sigma = \tau$.

*74. If E/F is a finite extension, then E/F is algebraic.

Theorem 28. *Let E/F be an extension field, and let $\alpha \in E$ be algebraic over F.*

(i) *There is a monic irreducible polynomial $p(x) \in F[x]$ having α as a root;*

(ii) *$p(x)$ is the monic polynomial of least degree in $F[x]$ having α as a root, hence is unique;*

(iii) *$F(\alpha) \cong F[x]/(p(x))$ by an isomorphism fixing F pointwise;*

(iv) *$[F(\alpha):F] = \partial p$.*

Proof. Choose $p(x)$ as the monic polynomial of least degree in $F[x]$ having α as a root ($p(x)$ exists because α is algebraic). The evaluation map $F[x] \to F(\alpha)$ taking $f(x) \mapsto f(\alpha)$ is surjective and has kernel $(p(x))$; the First Isomorphism Theorem gives an isomorphism $F[x]/(p(x)) \cong F(\alpha)$ (which fixes F). Since $F(\alpha)$ is a field, $(p(x))$ is a maximal, hence prime, ideal, and so $p(x)$ is irreducible. Finally, (iv) holds, by Theorem 27. $\square$

Definition. The polynomial $p(x)$ in Theorem 28 is called the **irreducible polynomial of α over F.**

If α is algebraic over F, then the description of $F(\alpha)$ as rational functions in α simplifies to $F(\alpha)$ as polynomials in α. In particular, the multiplicative inverse of $f(\alpha)$ is $a(\alpha)$, where $a(x)f(x) + b(x)p(x) = 1$ and $p(x)$ is the irreducible polynomial of α.

Corollary 29. *Let $\sigma: F \to F'$ be an isomorphism of fields, let $\sigma^*: F[x] \to F'[x]$ (defined by $\sum r_i x^i \mapsto \sum \sigma(r_i) x^i$) be the corresponding isomorphism of rings, let $p(x) \in F[x]$ be irreducible, and let $p^*(x) = \sigma^*(p(x)) \in F'[x]$. If β is a root of $p(x)$ and β' is a root of $p^*(x)$, then there is a unique isomorphism $\breve{\sigma}: F(\beta) \to F'(\beta')$ with $\breve{\sigma}(\beta) = \beta'$ and which extends σ.*

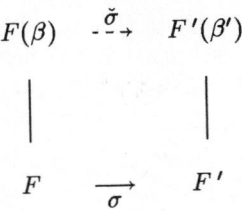

$$F(\beta) \quad \overset{\breve{\sigma}}{\dashrightarrow} \quad F'(\beta')$$

$$F \quad \underset{\sigma}{\longrightarrow} \quad F'$$

Proof. The isomorphism $\sigma^*: F[x] \to F'[x]$ carries the ideal $(p(x))$ onto the ideal $(p^*(x))$, and so the existence of $\breve{\sigma}$ follows from the theorem and Exercise 39; the uniqueness of $\breve{\sigma}$ follows from Exercise 73. (An alternative proof is to define $\breve{\sigma}: F(\beta) \to F'(\beta')$ by $\sum r_i \beta^i \mapsto \sum \sigma(r_i)(\beta')^i$ (of course, one must show that this is a well-defined isomorphism).) $\square$

Definition. A **splitting field** of $f(x) \in F[x]$ is a field extension E/F in which $f(x)$ splits (it is a product of linear factors) while $f(x)$ does not split in any proper subfield of E.

Example. If α is a primitive cube root of unity, then $x^3 - 1 \in \mathbb{Q}[x]$ splits over $\mathbb{C}$, but its splitting field is $\mathbb{Q}(\alpha)$.

Theorem 30. *Every polynomial $f(x) \in F[x]$ has a splitting field.*

Proof. By Kronecker's theorem (Theorem 17), there is an extension field K/F over which $f(x)$ splits. Define $E = F(\alpha_1, \ldots, \alpha_n)$, where $\alpha_1, \ldots, \alpha_n$ are the roots of $f(x)$ in K. It is plain that $f(x)$ splits over E, and $f(x)$ does not split over any proper subfield (which necessarily omits one of the α_i). $\square$

Lemma 31. *If $F \subset B \subset E$ are fields with $[E:B]$ and $[B:F]$ finite, then E/F is finite and*
$$[E:F] = [E:B][B:F].$$

Proof. Let $\{\alpha_1, \ldots, \alpha_m\}$ be a basis of E/B, and let $\{\beta_1, \ldots, \beta_n\}$ be a basis of B/F. It suffices to prove that $\{\beta_j \alpha_i : 1 \le i \le m, 1 \le j \le n\}$ is a basis of E/F.

This set spans. If $\gamma \in E$, then there are b_i in B with $\gamma = \sum b_i \alpha_i$. But each $b_i = \sum c_{ij} \beta_j$ for c_{ij} in F; hence $\gamma = \sum c_{ij} \beta_j \alpha_i$. To see that this set is independent, assume that $\sum c_{ij} \beta_j \alpha_i = 0$ for c_{ij} in F. Now $b_i = \sum c_{ij} \beta_j \in B$, so that independence of the α_i over B implies that

$b_i = 0$ for all i. Hence $\sum c_{ij}\beta_j = 0$ for all i, and so the independence of the β_j over F implies that $c_{ij} = 0$ for all i, j, as desired. □

Exercise

75. If $F \subset B \subset E$ are fields and E/F is finite, then both E/B and B/F are finite, and $[E:F] = [E:B][B:F]$.

Example. Let $E = \mathbb{Q}(\sqrt{2}, \sqrt{3})$ and let $F = \mathbb{Q}(\sqrt{2})$. Now $\sqrt{3}$ is algebraic over $\mathbb{Q}$ and its irreducible polynomial is $x^2 - 3$; it follows that $\sqrt{3}$ is algebraic over F. Moreover, the irreducible polynomial $p(x)$ of $\sqrt{3}$ over $\mathbb{Q}(\sqrt{2})$ is a divisor of $x^2 - 3$, so that $[E:\mathbb{Q}(\sqrt{2})] \leq 2$ (since $E = \mathbb{Q}(\sqrt{2}, \sqrt{3}) = \mathbb{Q}(\sqrt{2})(\sqrt{3})$). Indeed, $[E:\mathbb{Q}(\sqrt{2})] = 2$ because $\sqrt{3} \notin \mathbb{Q}(\sqrt{2})$ (there are no rationals a, b with $\sqrt{3} = a + b\sqrt{2}$).

Consider $\alpha = \sqrt{2} + \sqrt{3} \in E$. Note that α is algebraic over $\mathbb{Q}$ for $[E:\mathbb{Q}] = [E:\mathbb{Q}(\sqrt{2})][\mathbb{Q}(\sqrt{2}):\mathbb{Q}] = 4$; by Exercise 74, $E/\mathbb{Q}$ is an algebraic extension. What is the irreducible polynomial of α?

$$\alpha^2 = (\sqrt{2} + \sqrt{3})^2 = 5 + 2\sqrt{6}$$

and

$$\alpha^2 - 5 = 2\sqrt{6},$$

so that

$$\alpha^4 - 10\alpha^2 + 1 = 0.$$

The polynomial $p(x) = x^4 - 10x^2 + 1$ is irreducible over $\mathbb{Q}$, by Exercise 67. It follows that

$$E = \mathbb{Q}(\sqrt{2}, \sqrt{3}) = \mathbb{Q}(\sqrt{2} + \sqrt{3}) = \mathbb{Q}(\alpha),$$

for $4 = [E:\mathbb{Q}] = [E:\mathbb{Q}(\alpha)][\mathbb{Q}(\alpha):\mathbb{Q}] = 4[E:\mathbb{Q}(\alpha)]$; that is, E is a simple extension.

What is α^{-1}? Since $\alpha^4 - 10\alpha^2 + 1 = 0$, we have $\alpha(10\alpha - \alpha^3) = 1$ so that $\alpha^{-1} = 10\alpha - \alpha^3$. Replacing α by $\sqrt{2} + \sqrt{3}$, one can write α^{-1} explicitly in terms of $\sqrt{2}$ and $\sqrt{3}$.

The next result extends Corollary 29 so that it treats not necessarily irreducible polynomials. The second part of it introduces a new kind of polynomial.

Definition. Let $f(x) \in F[x]$ have the factorization into (not necessarily distinct) irreducibles:

$$f(x) = ap_1(x) \cdots p_t(x);$$

then $f(x)$ is **separable** if each $p_i(x)$ has no repeated roots.

Let F be a field and let $q(x) \in F[x]$ be irreducible. If the derivative $q'(x)$ is not the zero polynomial, then its degree is smaller than the degree of $q(x)$; hence $(q, q') = 1$ and $q(x)$ is separable, by Exercise 41. It follows that if F has characteristic 0, then every polynomial is separable; if F has characteristic p, then it is possible that $q' = 0$ (see the example below). Fields in which every nonconstant polynomial is separable are called **perfect**. (If E/F is an extension, then $\alpha \in E$ is called **separable** if either it is transcendental or its irreducible polynomial is separable; an extension is called **separable** if every one of its elements is separable.)

Example. Here is an example of an inseparable extension. Let $K = \mathbb{Z}_p(t)$, the field of all rational functions over $\mathbb{Z}_p$. The polynomial $q(x) = x^p - t$ is irreducible over K. Its splitting field E/K is not separable: if $\alpha \in E$ is a root of $q(x)$, then $x^p - t = (x - \alpha)^p$ in $E[x]$ because E has characteristic p. Note that $q'(x) = 0$.

Exercises

76. Show that a field F of characteristic p is perfect if and only if every element of F has a pth root in F.

77. Show that every finite field F is perfect. (Hint: The function $a \mapsto a^p$ is always an injection $F \to F$.)

Theorem 32. *Let $\sigma: F \to F'$ be an isomorphism of fields, let $f(x) \in F[x]$, and let $f^*(x) = \sigma^*(f(x))$ [3] be the corresponding polynomial in $F'[x]$; let E be a splitting field of $f(x)$ over F and let E' be a splitting field of $f^*(x)$ over F'.*

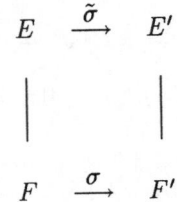

(i) *There is an isomorphism $\tilde{\sigma}: E \to E'$ extending σ.*

(ii) *If $f(x)$ is separable, then there are exactly $[E:F]$ extensions $\tilde{\sigma}$ of σ.*

Proof. (i) The proof is by induction on $[E:F]$. If $[E:F] = 1$, then $E = F$ and $f(x)$ is a product of linear factors in $F[x]$; it follows that $f^*(x)$ is also a product of linear factors, and so $E' = F'$; therefore, we may define $\tilde{\sigma} = \sigma$. If $[E:F] > 1$, choose an irreducible factor $p(x)$ of $f(x)$ having degree ≥ 2, and choose a root β of $p(x)$, hence a root of $f(x)$, which

---

[3]If $f(x) = \sum a_i x^i$, then $\sigma^*(f(x)) = \sum \sigma(a_i) x^i$.

must be in E. Let $p^*(x) \in F'[x]$ correspond to $p(x)$, and let $\beta' \in E'$ be a root of $p^*(x)$. By Corollary 29, for each such β' there is a unique isomorphism $\breve{\sigma}: F(\beta) \to F'(\beta')$ extending σ with $\breve{\sigma}(\beta) = \beta'$. Now E is a splitting field of $f(x)$ over $F(\beta)$ and E' is a splitting field of $f^*(x)$ over $F(\beta')$. Since $[E:F] = [E:F(\beta)][F(\beta):F]$, and since $[F(\beta):F] \geq 2$, it follows that $[E:F(\beta)] < [E:F]$. By induction, there exists $\tilde{\sigma}: E \to E'$ extending $\breve{\sigma}$, hence extending σ.

(ii) We modify the proof of (i), again proceeding by induction on $[E:F]$. If $[E:F] > 1$, let $f(x) = p(x)g(x)$, where $p(x)$ is irreducible of degree d, say. If $d = 1$, then we may replace $f(x)$ by $g(x)$ without changing the problem. If $d > 1$, choose a root β of $p(x)$. If $\tilde{\sigma}$ is any extension of σ to E, then $\sigma(\beta)$ is a root β' of $p^*(x)$; since $f^*(x)$ is separable, $p^*(x)$ has exactly d roots $\beta' \in E'$; by Corollary 29, there are exactly d isomorphisms $\breve{\sigma}: F(\beta) \to F'(\beta')$ extending σ, one for each β'. Now E is a splitting field of $f(x)$ over $F(\beta)$ and E' is a splitting field of $f^*(x)$ over $F'(\beta')$. Since $[E:F(\beta)] = [E:F]/d$, induction shows that each of the d isomorphisms $\breve{\sigma}$ has exactly $[E:F]/d$ extensions to E; therefore, σ has exactly $[E:F]$ extensions $\tilde{\sigma}$, because every τ extending σ has $\tau|F(\beta) = $ some $\breve{\sigma}$. $\square$

Corollary 33. *If $f(x) \in F[x]$, then any two splitting fields of $f(x)$ over F are isomorphic by an isomorphism fixing F pointwise.*

Proof. Choose $F = F'$ and σ the identity on F. $\square$

Corollary 34 (E.H. Moore). *Any two finite fields of order p^n are isomorphic.*

Proof. Any field F of order p^n is the splitting field of $x^q - x$ over $\mathbb{Z}_p$, where $q = p^n$. $\square$

One calls *the* field of order p^n the **Galois field** of this order and denotes it by $\mathrm{GF}(p^n)$, although $\mathrm{GF}(p)$ is usually denoted by $\mathbb{Z}_p$.

Exercise 64 asks for a construction of $\mathrm{GF}(8)$, with the hint to factor $x^8 - x = x^8 + x$ over $\mathbb{Z}_2$. Now

$$x^8 + x = x(x + 1)(x^3 + x + 1)(x^3 + x^2 + 1)$$

over $\mathbb{Z}_2$, and the two cubics are irreducible. Both $\mathbb{Z}_2[x]/(x^3 + x + 1)$ and $\mathbb{Z}_2[x]/(x^3 + x^2 + 1)$ are isomorphic because both are fields of order 8. Similarly, one sees that if $p(x)$ and $q(x)$ are irreducible polynomials over $\mathbb{Z}_p$ which have the same degree, then $\mathbb{Z}_p[x]/(p(x))$ and $\mathbb{Z}_p[x]/(q(x))$ are isomorphic.

Solvability by Radicals

Definition. A field extension B/F is a **pure extension** if $B = F(\alpha)$, where α^m lies in F for some positive integer m.

Definition. Let F be a field and let $f(x) \in F[x]$; then $f(x)$ is **solvable by radicals** over F if there is a tower of fields

$$F = B_0 \subset B_1 \subset \ldots \subset B_t,$$

where each B_{i+1}/B_i is a pure extension, and where B_t contains the splitting field E of $f(x)$ over F. One calls B_t/F a **radical extension.**

Exercises

*78. If $f(x)$ has degree n and is solvable by radicals over F, then there is a tower of pure extensions $F = B_0 \subset B_1 \subset \ldots \subset B_t$ with each $[B_{i+1} : B_i]$ prime.

*79. Let p be a prime and let F be a field containing all pth roots of unity. If $a \in F$, prove that $x^p - a$ either splits or is irreducible over F. (We give a fancy solution as Corollary 52.)

80. Let B/F be a finite extension. Prove that there is an extension E/B so that E/F is a splitting field of some polynomial $f(x) \in F[x]$. The smallest such extension E/F is called the **split closure** of B/F. (If $f(x)$ is a separable polynomial, then E/F is called the **normal closure** of B/F.)

*81. If E/F is a field extension and B and C are intermediate fields ($F \subset B \subset E$ and $F \subset C \subset E$), then their **compositum** $B \vee C$ is the subfield of E generated by B and C; that is, the intersection of all the subfields of E containing B and C. Show that if E/F is the split closure of B/F, then $E = B_1 \vee \ldots \vee B_r$, where each B_i is isomorphic to B via an isomorphism which fixes F.

*82. Using the preceding exercise, prove that the split closure of a radical extension is itself a radical extension. Conclude that, in the definition of solvable by radicals, one can assume that the last field B_t is a splitting field of some polynomial over F.

We illustrate the definition of solvability by radicals by showing that quadratics, cubics, and quartics over subfields F of $\mathbb{C}$ are solvable by radicals (the formulas are not true for arbitrary fields F; for example, the quadratic formula cannot hold when F has characteristic 2).

If $f(x) = x^2 + bx + c \in \mathbb{Q}[x]$, define $F = \mathbb{Q}(b, c)$ and $B = F(\sqrt{b^2 - 4c})$. Then B/F is a pure extension, and B is the splitting field of $f(x)$ over F; therefore, $f(x)$ is solvable by radicals over F.

If $f(x) = x^3 + qx + r$, define $F = \mathbb{Q}(q, r)$, define

$$B_1 = F(\sqrt{r^2 + 4q^3/27}),$$

and define $B_2 = B_1(y)$, where $y^3 = \frac{1}{2}(-r + \sqrt{r^2 + 4q^3/27})$. Define $B_3 = B_2(z)$, where

$$z^3 = \frac{1}{2}\left(-r - \sqrt{r^2 + 4q^3/27}\right).$$

The cubic formula says that the roots of $f(x)$ are of the form $\omega y + \omega' z$, where ω, ω' are cube roots of unity and $(\omega y)(\omega' z) = -q/3$. Therefore, if we define $B_4 = B_3(\omega)$, then the splitting field E of $f(x)$ is contained in B_4, and $f(x)$ is solvable by radicals. Note that it is possible that E is a proper subfield of B_4, for E need not contain ω; for example, $f(x)$ may have three real roots and E is a subfield of $\mathbb{R}$.

If $f(x) = x^4 + qx^2 + rx + s$, define $F = \mathbb{Q}(q, r, s)$. In the discussion of the quartic formula, we saw that it suffices to find three numbers k, ℓ, and m. Now k^2 is a root of a certain cubic polynomial in $F[x]$, so that there is a tower of pure field extensions $F \subset B_1 \subset \ldots \subset B_4$ with $k^2 \in B_4$. Define $B_5 = B_4(k)$. Since $2m = k^2 + q + r/k$ and $2\ell = k^2 + q - r/k$, B_5 contains ℓ and m. The quartic formula gives the roots of $f(x)$ as the roots of

$$(x^2 + kx + \ell)(x^2 - kx + m),$$

so that the tower of pure extensions can be lengthened two steps, by adjoining $\sqrt{k^2 - 4\ell}$ and $\sqrt{k^2 - 4m}$, with the last extension, B_7, containing the splitting field of $f(x)$. Therefore, $f(x)$ is solvable by radicals.

It should be plain that, conversely, if a polynomial $f(x)$ is solvable by radicals, then there is an expression for its roots in terms of the coefficients, the field operations, and extraction of roots.

The Galois Group

The next lemma, though very easy to prove, is fundamental.

Lemma 35. *Let $f(x) \in F[x]$ and let E/F be the splitting field of $f(x)$ over F. If $\sigma: E \to E$ is an **automorphism** (an isomorphism of E with itself) fixing F pointwise and if α is a root of $f(x)$, then $\sigma(\alpha)$ is also a root of $f(x)$.*

Proof. Let $f(x) = a_0 + a_1 x + \cdots + a_n x^n$, so that $a_0 + a_1\alpha + \cdots + a_n\alpha^n = 0$. Applying σ gives $\sigma(a_0) + \sigma(a_1)\sigma(\alpha) + \cdots + \sigma(a_n)\sigma(\alpha)^n = a_0 + a_1\sigma(\alpha) + \cdots + a_n\sigma(\alpha)^n = 0$, because σ fixes F. Therefore, $\sigma(\alpha)$ is a root of $f(x)$. $\square$

Definition.[4] Let E/F be a field extension. Its **Galois group**, denoted by $\mathrm{Gal}(E/F)$, is the set of all the automorphisms of E that fix F (pointwise)

---

[4]This is not the definition of Galois; it is the modern version introduced by E. Artin and isomorphic to the original version. The defect of this definition is that it does not seem to arise naturally; see Appendix 4.

under the binary operation of composition. If $f(x) \in F[x]$ has splitting field E, then the **Galois group** of $f(x)$ is $\mathrm{Gal}(E/F)$.

It is easy to check that $\mathrm{Gal}(E/F)$ is a group.

Theorem 36. *If $f(x) \in F[x]$ has n distinct roots in its splitting field E, then $\mathrm{Gal}(E/F)$ is isomorphic to a subgroup of the symmetric group S_n.*

Proof. Let $X = \{\alpha_1, \ldots, \alpha_n\}$ be the set of all the roots of $f(x)$ in E. By Lemma 35, if $\sigma \in \mathrm{Gal}(E/F)$, then $\sigma(X) = X$. The map $\mathrm{Gal}(E/F) \to S_X$ defined by $\sigma \mapsto \sigma|X$ is easily seen to be a homomorphism; it is an injection, by Exercise 73. Finally, $S_X \cong S_n$.   □

For example, the Galois group of a quartic polynomial is a subgroup of S_4, and the Galois group of a quintic polynomial is a subgroup of S_5.

Theorem 37. *If $f(x) \in F[x]$ is a separable polynomial, and if E/F is its splitting field, then $|\mathrm{Gal}(E/F)| = [E:F]$.*

Proof. By Theorem 32(ii) with $F = F'$, $E = E'$, and $\sigma: F \to F$ the identity, there are exactly $[E:F]$ automorphisms of E that fix F.   □

Examples

1. The splitting field of $x^2 + 1$ over $\mathbb{R}$ is, of course, $\mathbb{C}$, and $|\mathrm{Gal}(\mathbb{C}/\mathbb{R})| \leq 2$, by Theorem 36. In fact, $|\mathrm{Gal}(\mathbb{C}/\mathbb{R})| = 2$ because the group contains the automorphism

$$\sigma: z = a + ib \mapsto \bar{z} = a - ib.$$

Notice that $\sigma: i \mapsto -i$ and $-i \mapsto i$, and so it interchanges the roots. One sees that the elements of the Galois group should be regarded as generalizations of complex conjugation.

2. Let $f(x) = x^3 - 1 \in \mathbb{Q}[x]$; $f(x)$ is separable, because $\mathbb{Q}$ has characteristic 0. Now $f(x) = (x - 1)(x^2 + x + 1)$ is a factorization of $f(x)$ into irreducibles. If E is the splitting field of $f(x)$ over $\mathbb{Q}$, then $E = \mathbb{Q}(\omega)$, where ω is a primitive cube root of unity, that is, ω is a root of $x^2 + x + 1$. Since $2 = [E:\mathbb{Q}] = |\mathrm{Gal}(E/\mathbb{Q})|$, by Theorem 37, the Galois group is cyclic of order 2.

3. Let $g(x) = x^3 - 2 \in \mathbb{Q}[x]$. The splitting field of $f(x)$ is $\mathbb{Q}(\alpha, \omega)$, where α is the real cube root of 2 and ω is a complex cube root of unity. Since $g(x)$ is irreducible over $\mathbb{Q}$, we have $[\mathbb{Q}(\alpha):\mathbb{Q}] = 3$. But $\mathbb{Q}(\alpha)$ consists wholly of real numbers, and so it cannot be the splitting field E of $g(x)$. Hence

$$|\mathrm{Gal}(E/\mathbb{Q})| = [E:\mathbb{Q}] = [E:\mathbb{Q}(\alpha)][\mathbb{Q}(\alpha):\mathbb{Q}] = 3[E:\mathbb{Q}(\alpha)] > 3;$$

it follows that $\mathrm{Gal}(E/\mathbb{Q}) \cong S_3$, by Theorem 36.

Exercises

*83. Let p be a prime and let F be a field containing all the pth roots of
unity. If E/F is a pure extension of degree p, then $\text{Gal}(E/F)$ is cyclic
of order p.

*84. Let $f(x) \in F[x]$, let E/F be a splitting field, and let $G = \text{Gal}(E/F)$
be the Galois group.

  (i) If $f(x)$ is irreducible, then G acts **transitively** on the set of all
  roots of $f(x)$ (if α and β are any two roots of $f(x)$ in E, there
  exists $\sigma \in G$ with $\sigma(\alpha) = \beta$). (Hint: Theorem 32(i).)

  (ii) If $f(x)$ has no repeated roots and G acts transitively on the
  roots, then $f(x)$ is irreducible. (Hint: If $f(x) = g(x)h(x)$, then
  the gcd $(g(x), h(x)) = 1$; if α is a root of $g(x)$ such that $\sigma(\alpha)$ is
  a root of $h(x)$, then $\sigma(\alpha)$ is a common root of $g(x)$ and $h(x)$.)

Lemma 38. *Let $F \subset B \subset E$ be a tower of fields with B/F the splitting field
of some polynomial $f(x) \in F[x]$. If $\sigma \in \text{Gal}(E/F)$, then $\sigma|B \in \text{Gal}(B/F)$.*

Proof. It suffices to prove that $\sigma(B) = B$. If $\alpha_1, \ldots, \alpha_n$ are the roots of
$f(x)$, then $B = F(\alpha_1, \ldots, \alpha_n)$. Now $\sigma(F) = F$, and $\sigma(\alpha_i) \in B$ for all i; it
follows from Exercise 73 that $\sigma(B) = B$, as desired.   □

Theorem 39. *Let $F \subset B \subset E$ be a tower of fields with B/F the splitting
field of some polynomial $f(x) \in F[x]$ and E/F the splitting field of some
$g(x) \in F[x]$. Then $\text{Gal}(E/B)$ is a normal subgroup of $\text{Gal}(E/F)$, and*

$$\text{Gal}(E/F)/\text{Gal}(E/B) \cong \text{Gal}(B/F).$$

Proof. Define $\psi: \text{Gal}(E/F) \to \text{Gal}(B/F)$ by $\sigma \mapsto \sigma|B$; Lemma 38 says
that ψ does take its values in $\text{Gal}(B/F)$. It is easily seen that ψ is a homomorphism with kernel $\text{Gal}(E/B)$, so that the latter is a normal subgroup
of $\text{Gal}(E/F)$. If $\tau \in \text{Gal}(B/F)$, then Theorem 32 shows that there is an
automorphism $\tilde{\tau}$ of E with $\psi(\tilde{\tau}) = \tilde{\tau}|B = \tau$. Hence ψ is surjective, and the
First Isomorphism Theorem (Theorem A5) gives the result.   □

Remark. The hypothesis that E/F is a splitting field enters only in showing that ψ is surjective. Without this hypothesis, one can prove only that
the quotient group is isomorphic to a subgroup of $\text{Gal}(B/F)$.

Recall that a finite group is *solvable* if it has a normal series with abelian
factor groups; moreover (Theorem A20), every quotient and every subgroup
of a solvable group is itself solvable.

Theorem 40. *Let $f(x) \in F[x]$ have degree n, and assume that F contains all the pth roots of unity for all primes p dividing $n!$. If $f(x)$ is solvable by radicals, then its Galois group is a solvable group.*

Proof. Since $f(x)$ is solvable by radicals, there is a radical extension $F = B_0 \subset B_1 \subset \ldots \subset B_t$ with $E \subset B_t$, where E is the splitting field of $f(x)$ over F. By Exercise 78, we may assume each $[B_{i+1}:B_i]$ is prime. Define $G_i = \mathrm{Gal}(B_t/B_i)$. By Exercise 79, each B_{i+1} is a splitting field of a polynomial over B_i, and so Theorem 39 shows that

$$G = G_0 \supset G_1 \supset \ldots \supset G_t = \{1\}$$

is a normal series; moreover, $\mathrm{Gal}(B_t/B_i)/\mathrm{Gal}(B_t/B_{i+1}) \cong \mathrm{Gal}(B_{i+1}/B_i)$, and this last group is cyclic of prime order, by Exercise 83. Therefore, G is a solvable group. Finally, Exercise 82 allows us to assume that B_t is a splitting field of a polynomial over F, so another application of Theorem 39 shows that $\mathrm{Gal}(E/F)$ is a quotient group of the solvable group $G = \mathrm{Gal}(B_t/F)$; hence it, too, is solvable, by Theorem A20. $\square$

Primitive Roots of Unity

The hypothesis in Theorem 40 that F contain certain roots of unity can be dropped, but we give a preliminary discussion from group theory before proving this.

Lemma 41. *If $C = \langle a \rangle$ is a cyclic group of order n and generator a, then C has a unique subgroup of order d for each divisor d of n.*

Proof. If $n = dc$, we show that a^c has order d (and so $\langle a^c \rangle$ is a subgroup of order d). Clearly $(a^c)^d = 1$; we claim that d is the smallest such power. If $(a^c)^r = 1$, then $n|cr$; hence $cr = ns = dcs$ (for some integer s) and $r = ds \geq d$. To prove uniqueness, assume that $\langle x \rangle$ is a subgroup of order d (recall that every subgroup of a cyclic group is cyclic, by Lemma A1). Now $x = a^m$ and $1 = x^d = a^{md}$; hence $md = nk$ for some integer k. Therefore, $x = a^m = (a^{n/d})^k = (a^c)^k$, so that $\langle x \rangle \subset \langle a^c \rangle$. Since both subgroups have the same order d, it follows that $\langle x \rangle = \langle a^c \rangle$. $\square$

Recall Theorem A2: if C is a cyclic group with generator x and order n, then x^k is also a generator of C if and only if k and n are relatively prime; it follows that if $g(C)$ denotes the set of all generators of C, then $|g(C)| = \varphi(n)$, where φ is the Euler φ-function ($\varphi(1) = 1$ and, if $n > 1$, then $\varphi(n) = |\{k \in \mathbb{Z} : 1 \leq k < n \text{ and } (k,n) = 1\}|$.)

Theorem 42. *If n is a positive integer, then*

$$n = \sum_{\substack{d|n \\ 1 \leq d \leq n}} \varphi(d).$$

Proof. If G is a group, then it is easy to see that it is the disjoint union

$$G = \uplus g(C),$$

where C ranges over all the cyclic subgroups of G. If G is a cyclic group of order n, then counting gives

$$n = \sum |g(C)| = \sum \varphi(d),$$

where the summation ranges over all divisors d of n, and each d occurs with multiplicity the number of cyclic subgroups of order d. By the lemma, each d occurs exactly once.   □

Theorem 43. *A group G of order n is cyclic if and only if, for each divisor d of n, there is at most one cyclic subgroup of order d.*

Proof. If G is cyclic, then the result follows from Lemma 41. Conversely, write G as a disjoint union (as in the preceding proof): $G = \uplus g(C)$. Hence $n = |G| = \sum |g(C)|$, where the summation is over all cyclic subgroups C of G. Since G has at most one cyclic subgroup of order d, Theorem 42 gives

$$n = \sum |g(C)| \leq \sum \varphi(d) = n.$$

Therefore, G has exactly one cyclic subgroup of order d, for every divisor d of n; in particular, there is a cyclic subgroup of order n, and G is cyclic. □

Theorem 44. *If F is a field with multiplicative group $F^{\#} = F - \{0\}$, then every finite subgroup G of $F^{\#}$ is cyclic.*

Proof. Suppose $|G| = n$ and $d \mid n$. If C is a cyclic subgroup of G of order d, then Lagrange's theorem gives $x^d = 1$ for each of the d elements $x \in C$. Were there a second cyclic subgroup of order d, then G would contain at least $d + 1$ elements x with $x^d = 1$. But the polynomial $x^d - 1$ has at most d roots in a field, and so G has at most one cyclic subgroup of order d. Theorem 43 now shows that G is cyclic.   □

Corollary 45. *If n if a fixed positive integer, then all the nth roots of unity in a field F form a cyclic multiplicative group.*

Corollary 46. *If F is a finite field, then $F^{\#}$ is cyclic and $F = \mathbb{Z}_p(\alpha)$ for some α.*

Definition. A generator of $F^{\#}$ when F is finite is called a **primitive element**.

Lemma 47. *If α is a primitive element of $GF(p^n)$, then α is a root of an irreducible polynomial of degree n.*

Proof. If the irreducible polynomial of α over $\mathbb{Z}_p$ has degree d, then $\mathbb{Z}_p(\alpha)$ has order p^d. But this subfield is all of $GF(p^n)$ because α is a primitive element; hence $d = n$. $\square$

Theorem 48. $\mathrm{Gal}(GF(p^n)/GF(p)) \cong \mathbb{Z}_n$ *with generator* $u \mapsto u^p$.

Remark. This generator is called the **Frobenius automorphism**.

Proof. Denote $GF(p^n)$ by K and denote the Galois group by G. If α is a primitive element, then its irreducible polynomial $p(x)$ has degree n (Lemma 47), and so K contains at most n of its roots. If $\sigma \in G$, then σ is completely determined by $\sigma(\alpha)$ (because every nonzero element of K has the form α^i and $\sigma(\alpha^i) = \sigma(\alpha)^i$). But $\sigma(\alpha)$ is a root of $p(x)$, by Lemma 35; it follows that $|G| \leq n$. On the other hand, $\sigma\colon u \mapsto u^p$ does lie in G; moreover, if $j < n$, then $\sigma^j \neq 1$ (otherwise $u^{p^j} = u$ for all u, and K would contain p^n roots of $x^{p^j} - x$, a contradiction). The theorem follows. $\square$

Lemma 49. *Let n be a positive integer and let F be a field. If the characteristic of F is either 0 or is a prime not dividing n, then $x^n - 1$ has n distinct roots in a splitting field.*

Proof. If $f(x) = x^n - 1$, then its derivative $f'(x) = nx^{n-1}$. By hypothesis, this is not zero, and so the gcd $(f, f') = 1$; therefore, $f(x)$ has no repeated roots. $\square$

One must say something about characteristic p, for then

$$x^p - 1 = (x - 1)^p.$$

Definition. Let n be a fixed positive integer and let F be a field. A generator of the group of all nth roots of unity is called a **primitive root of unity** if the characteristic of F is either 0 or a prime not dividing n.

A primitive nth root of unity in $\mathbb{C}$ is $e^{2\pi i/n}$.

Theorem 50. *If F is a field and $E = F(\alpha)$, where α is a primitive nth root of unity, then $\mathrm{Gal}(E/F)$ is abelian.*

Proof. Note that E is the splitting field of $x^n - 1$ because α is a primitive nth root of unity. Now $\sigma(\alpha) = \alpha^i$ for every $\sigma \in \mathrm{Gal}(E/F)$; moreover, Theorem A2(ii) says i must be relatively prime to n, for $\sigma|\langle\alpha\rangle$ is an automorphism of $\langle\alpha\rangle$. Define $\psi\colon \mathrm{Gal}(E/F) \to (\mathbb{Z}/n\mathbb{Z})^{\#}$ (the multiplicative group of all congruence classes of integers relatively prime to n) by $\sigma \mapsto \bar{i}$, where $\sigma(\alpha) = \alpha^i$. It is routine to check that ψ is a homomorphism; it is injective, by Exercise 73. Therefore $\mathrm{Gal}(E/F)$ is isomorphic to a subgroup of an abelian group, hence it is abelian. $\square$

Note that $(\mathbb{Z}/n\mathbb{Z})^{\#}$ need not be cyclic; for example, $(\mathbb{Z}/8\mathbb{Z})^{\#}$ consists of the congruence classes of 1, 3, 5, 7, and it is isomorphic to the 4-group.

(There is a deep partial converse of Theorem 50. The Kronecker–Weber Theorem states that every finite abelian extension of $\mathbb{Q}$ (that is, a finite extension $E/\mathbb{Q}$ with $\mathrm{Gal}(E/\mathbb{Q})$ abelian) can be imbedded in a cyclotomic extension $\mathbb{Q}(\omega)$, where ω is some root of unity.)

Example. If p is a prime, then $\zeta = e^{2\pi i/p}$ is a primitive pth root of unity over $\mathbb{Q}$. As in the proof of Theorem 50, $\mathbb{Q}(\zeta)$ is the splitting field of $x^p - 1$ over $\mathbb{Q}$. But $x^p - 1 = (x - 1)\Phi_p(x)$, where $\Phi_p(x)$ is the pth cyclotomic polynomial; since $\Phi_p(x)$ is irreducible over $\mathbb{Q}$ (Corollary 25), we have $|\mathrm{Gal}(\mathbb{Q}(\zeta)/\mathbb{Q})| = p - 1$. Theorem 50 now gives

$$\mathrm{Gal}(\mathbb{Q}(\zeta)/\mathbb{Q}) \cong (\mathbb{Z}_p)^{\#};$$

by Corollary 46, the latter group is cyclic, being the multiplicative group of a field.

Let us use a variant of Theorem 50 to give a sophisticated solution to Exercise 79.

Theorem 51. *Let F contain a primitive nth root of unity, and let $f(x) = x^n - a$. If E/F is a splitting field of $f(x)$, then restriction gives an injection*

$$G = \mathrm{Gal}(E/F) \to \mathbb{Z}_n.$$

Moreover, $f(x)$ is irreducible if and only if this map is surjective.

Proof. If ω is a primitive nth root of unity and if α is a root of $f(x)$, then the list of all the roots of $f(x)$ is: $\alpha, \alpha\omega, \ldots, \alpha\omega^{n-1}$. If $\sigma \in G$, then $\sigma(\alpha) = \alpha\omega^i$, and σ is completely determined by i; it follows easily that $\sigma \mapsto \bar{i}$ is the advertised injection. Now this map is surjective if and only if G acts transitively on the roots of $f(x)$. By Exercise 84, this is equivalent to the irreducibility of $f(x)$. $\quad\square$

Corollary 52. *Let p be a prime and let F be a field containing a primitive pth root of unity. If $a \in F$, then $x^p - a$ either splits or is irreducible.*

Proof. Consider the map $\mathrm{Gal}(E/F) \to \mathbb{Z}_p$ of the theorem. If $f(x)$ splits, then its image is trivial; if $f(x)$ does not split, then its image is a nontrivial subgroup of $\mathbb{Z}_p$. But $\mathbb{Z}_p$ has no proper nontrivial subgroups, so that the map must be surjective and $f(x)$ is irreducible. $\quad\square$

Insolvability of the Quintic

Recall Theorem A21: If G is a group having a solvable normal subgroup H such that G/H is solvable, then G is solvable. Here is the improved version of Theorem 40 which needs no assumption about roots of unity.

Theorem 53. *Let $f(x) \in F[x]$ be solvable by radicals over a field F, and let E/F be its splitting field. Then $\mathrm{Gal}(E/F)$ is a solvable group.*

Proof. By hypothesis, there is a radical extension

$$F = B_0 \subset B_1 \subset \ldots \subset B_t,$$

with $E \subset B_t$. Only finitely many roots of unity have been adjoined to F, say, the k_1th, k_2th,$\ldots$, k_sth roots of unity. If k is the product of all the k_i, then there exists a primitive kth root of unity, say, α (this is surely true if F has characteristic 0; if F has characteristic p, then we may assume that no k_i is divisible by p because the only pth root of unity is 1). Adjust the original tower by adjoining α first:

$$F = B_0 \subset F(\alpha) \subset B_1(\alpha) \subset \ldots \subset B_t(\alpha) = B'.$$

Notice that each extension in this tower is pure and that $E \subset B_t(\alpha)$. Since $F(\alpha)/F$ is a splitting field, the remark after Theorem 39 shows that $\mathrm{Gal}(B'/F)/\mathrm{Gal}(B'/F(\alpha))$ is isomorphic to a subgroup of the abelian group $\mathrm{Gal}(F(\alpha)/F)$; hence it is itself abelian, by Theorem 50. The group $\mathrm{Gal}(B'/F)$ thus has a solvable normal subgroup $\mathrm{Gal}(B'/F(\alpha))$, by Theorem 40, with an abelian, hence solvable quotient; it follows from Theorem A21 that $\mathrm{Gal}(B'/F)$ is solvable. Finally, Exercise 82 allows us to assume that B'/F is a splitting field, so that Theorem 39 applies to show that $\mathrm{Gal}(E/F)$ is a quotient group of the solvable group $\mathrm{Gal}(B'/F)$ and hence is solvable, by Theorem A20. □

Theorem 54 (Abel–Ruffini). *There exists a quintic polynomial $f(x) \in \mathbb{Q}[x]$ that is not solvable by radicals.*

Proof. Let $f(x) = x^5 - 4x + 2$; $f(x)$ is irreducible over $\mathbb{Q}$, by Eisenstein's criterion. Let $E/\mathbb{Q}$ be the splitting field of $f(x)$ contained in $\mathbb{C}$,[5] and let $G = \mathrm{Gal}(E/\mathbb{Q})$. If α is a root of $f(x)$, then $[\mathbb{Q}(\alpha):\mathbb{Q}] = 5$, and so $|G| = [E:\mathbb{Q}(\alpha)][\mathbb{Q}(\alpha):\mathbb{Q}]$ is divisible by 5. We now use elementary calculus; $f(x)$ has exactly two critical points, namely, $\pm\sqrt[4]{4/5} \sim \pm.946$, and $f(\sqrt[4]{4/5}) < 0$ and $f(-\sqrt[4]{4/5}) > 0$; it follows easily that $f(x)$ has exactly three real roots. Regarding G as a group of permutations on the 5 roots, we note that G contains a 5-cycle (the only kind of element of order 5 in S_5) and σ, the restriction of complex conjugation, which interchanges the two complex roots while fixing the three real roots (so σ is a transposition). By Theorem A37, S_5 is generated by any transposition and any 5-cycle, so that $G \cong S_5$; hence G is not a solvable group, by Theorem A32, and Theorem 53 shows that $f(x)$ is not solvable by radicals. □

---

[5] We are assuming the fundamental theorem of algebra (Theorem 71).

Independence of Characters

Definition. A **character** of a group G in a field E is a homomorphism $\sigma\colon G \to E^{\#}$, where $E^{\#} = E - \{0\}$ is the multiplicative group of E.

Definition. A set $\{\sigma_1, \ldots, \sigma_n\}$ of characters of a group G in a field E is **independent** if there does not exist a nonzero "vector" $(a_i) \in E^n$ with

$$\sum a_i \sigma_i(x) = 0 \quad \text{for all } x \in G.$$

Lemma 55 (Dedekind). *Every set $\{\sigma_1, \ldots, \sigma_n\}$ of distinct characters of a group G in a field E is independent.*

Proof. The proof is by induction on n. If $n = 1$, then $a_1 \sigma_1(x) = 0$ implies that $a_1 = 0$ because $\sigma_1(x) \neq 0$. Let $n > 1$ and assume there is an equation

$$(1) \qquad a_1 \sigma_1(x) + \cdots + a_n \sigma_n(x) = 0 \quad \text{for all } x \in G,$$

where not all $a_i = 0$. We may assume that every $a_i \neq 0$ lest induction apply; multiplying by a_n^{-1} if necessary, we may further assume that $a_n = 1$. Since $\sigma_n \neq \sigma_1$, there exists $y \in G$ with $\sigma_n(y) \neq \sigma_1(y)$. In Eq. (1), replace x by yx to obtain

$$a_1 \sigma_1(y)\sigma_1(x) + \cdots + a_{n-1}\sigma_{n-1}(y)\sigma_{n-1}(x) + \sigma_n(y)\sigma_n(x) = 0.$$

Multiply by $\sigma_n(y)^{-1}$ to obtain an equation

$$a_1 \sigma_n(y)^{-1}\sigma_1(y)\sigma_1(x) + \cdots + \sigma_n(x) = 0;$$

subtract this from Eq. (1) to obtain a sum of $n - 1$ terms

$$a_1 [1 - \sigma_n(y)^{-1}\sigma_1(y)]\sigma_1(x) + \cdots = 0.$$

By induction, each of the coefficients is 0. Since $a_1 \neq 0$, we have $1 = \sigma_n(y)^{-1}\sigma_1(y)$; hence $\sigma_n(y) = \sigma_1(y)$, a contradiction. $\quad\square$

Corollary 56. *Every set $\{\sigma_1, \ldots, \sigma_n\}$ of distinct automorphisms of a field E is independent.*

Proof. An automorphism σ of E restricts to a (group) homomorphism $\sigma\colon E^{\#} \to E^{\#}$, hence is a character. $\quad\square$

Definition. Let $\mathrm{Aut}(E)$ be the group of all the automorphisms of a field E. If G is a subset of $\mathrm{Aut}(E)$, then

$$E^G = \{\alpha \in E \colon \sigma(\alpha) = \alpha \ \text{ for all } \ \sigma \in G\}$$

is called the **fixed field**.

It is easy to see that E^G is a subfield of E. The most important instance of this definition is when G is a subgroup of $\mathrm{Aut}(E)$, but there is an application when G is only a subset. Note that

$$H \subset G \quad \text{implies} \quad E^G \subset E^H.$$

Examples

1. If E/F is a field extension with Galois group $G = \mathrm{Gal}(E/F)$, then

$$F \subset E^G \subset E;$$

we shall presently consider whether E^G/F is a proper extension.

2. Let $E = F(x_1, \ldots, x_n)$ be the rational functions in several variables over a field F. Then $G \cong S_n$ can be regarded as a subgroup of $\mathrm{Aut}(E)$; it acts by permuting the variables. The elements of the fixed field E^G are called **symmetric functions**[6] over F.

Lemma 57. *If $G = \{\sigma_1, \ldots \sigma_n\}$ is a set of automorphisms of E, then*

$$[E \colon E^G] \geq n.$$

Proof. Otherwise $[E \colon E^G] = r < n$; let $\{\alpha_1, \ldots, \alpha_r\}$ be a basis of E/E^G. Consider the linear system over E of r equations in n unknowns:

$$\sigma_1(\alpha_1)x_1 + \cdots + \sigma_n(\alpha_1)x_n = 0$$
$$\sigma_1(\alpha_2)x_1 + \cdots + \sigma_n(\alpha_2)x_n = 0$$
$$\cdots\cdots\cdots\cdots\cdots$$
$$\sigma_1(\alpha_r)x_1 + \cdots + \sigma_n(\alpha_r)x_n = 0.$$

Since $r < n$, there is a nontrivial solution $(x_1, \ldots, x_n)$. For any $\beta \in E$, we have $\beta = \sum b_i \alpha_i$ where $b_i \in E^G$. Multiply the ith row of the system by $\sigma_1(b_i)$ to obtain the system with ith row:

$$\sigma_1(b_i)\sigma_1(\alpha_i)x_1 + \cdots + \sigma_1(b_i)\sigma_n(\alpha_i)x_n = 0.$$

But $\sigma_1(b_i) = b_i = \sigma_j(b_i)$ for all i, j because $b_i \in E^G$. The system thus has ith row:

$$\sigma_1(b_i\alpha_i)x_1 + \cdots + \sigma_n(b_i\alpha_i)x_n = 0.$$

Now add to get

$$\sigma_1(\beta)x_1 + \cdots + \sigma_n(\beta)x_n = 0.$$

---

[6]Symmetric functions arise naturally: if $f(x) = \prod(x - \alpha_i) = x^n + s_{n-1}x^{n-1} + \cdots + s_1 x + s_0$, then each of the coefficients s_j is a symmetric function of the roots $\alpha_1, \ldots, \alpha_n$. This observation is the starting point of Lagrange and Galois (see Appendix 4).

As β is an arbitrary element of E, the independence of the characters $\{\sigma_1, \ldots, \sigma_n\}$ is violated. This contradiction proves the theorem. $\quad\square$

Theorem 58. *If* $G = \{\sigma_1, \ldots, \sigma_n\}$ *is a subgroup of* $\mathrm{Aut}(E)$, *then*

$$[E : E^G] = |G|.$$

Proof. It suffices to prove $[E : E^G] \leq |G|$. Otherwise $[E : E^G] > n$; let $\{\omega_1, \ldots, \omega_{n+1}\}$ be linearly independent vectors in E over E^G. Consider the system of n equations in $n + 1$ unknowns:

$$\sigma_1(\omega_1)x_1 + \cdots + \sigma_1(\omega_{n+1})x_{n+1} = 0$$
$$\cdots\cdots\cdots\cdots\cdots$$
$$\sigma_n(\omega_1)x_1 + \cdots + \sigma_n(\omega_{n+1})x_{n+1} = 0.$$

There is a nontrivial solution $(x_1, \ldots, x_{n+1})$ over E; we proceed to normalize it. Choose a solution having the least number r of nonzero components, say $(a_1, \ldots, a_r, 0, \ldots, 0)$; by reindexing the ω_i, we may assume that all nonzero components come first. Note that $r \neq 1$ lest $\sigma_1(\omega_1)a_1 = 0$ imply $a_1 = 0$. Multiplying by its inverse if necessary, we may assume that $a_r = 1$. Not all $a_i \in E^G$ lest the row corresponding to the identity of G violate the linear independence of $\{\omega_1, \ldots, \omega_{n+1}\}$. Our last assumption is that a_1 does not lie in E^G (this, too, can be accomplished by reindexing the ω_i). There thus exists σ_k with $\sigma_k(a_1) \neq a_1$. The original system has jth row

(1) $\qquad\qquad \sigma_j(\omega_1)a_1 + \cdots + \sigma_j(\omega_{r-1})a_{r-1} + \sigma_j(\omega_r) = 0.$

Apply σ_k to this system to obtain

$$\sigma_k\sigma_j(\omega_1)\sigma_k(a_1) + \cdots + \sigma_k\sigma_j(\omega_{r-1})\sigma_k(a_{r-1}) + \sigma_k\sigma_j(\omega_r) = 0.$$

Since G is a group, $\sigma_k\sigma_1, \ldots, \sigma_k\sigma_n$ is just a permutation of $\sigma_1, \ldots, \sigma_n$. Setting $\sigma_k\sigma_j = \sigma_i$, the system has ith row

$$\sigma_i(\omega_1)\sigma_k(a_1) + \cdots + \sigma_i(\omega_{r-1})\sigma_k(a_{r-1}) + \sigma_i(\omega_r) = 0.$$

Subtract this from the ith row of Eq. (1) to obtain a new system with ith row:

$$\sigma_i(\omega_1)[a_1 - \sigma_k(a_1)] + \cdots + \sigma_i(\omega_{r-1})[a_{r-1} - \sigma_k(a_{r-1})] = 0.$$

Since $a_1 - \sigma_k(a_1) \neq 0$, we have found a nontrivial solution of the original system having less than r nonzero components, a contradiction. $\quad\square$

Corollary 59. *If* G, H *are subgroups of* $\mathrm{Aut}(E)$ *with* $E^G = E^H$, *then* $G = H$.

Proof. If $\sigma \in G$, then clearly σ fixes E^G. To prove the converse, suppose σ fixes E^G and $\sigma \notin G$. Then E^G is fixed by the $n+1$ elements in $G \cup \{\sigma\}$, so Lemma 57 and Theorem 58 give the contradiction:

$$n = |G| = [E : E^G] \geq [E : E^{G \cup \{\sigma\}}] \geq n + 1.$$

Therefore, if σ fixes E^G, then $\sigma \in G$.

If $\sigma \in G$, then σ fixes $E^G = E^H$, and hence $\sigma \in H$; the reverse inclusion is proved the same way. $\square$

Galois Extensions

The discussion of Galois groups began with a *pair* of fields, namely, an extension E/F that is a splitting field of some polynomial $f(x) \in F[x]$. Suppose that $G = \mathrm{Gal}(E/F)$; it is easy to see that

$$F \subset E^G \subset E.$$

A natural question is whether $F = E^G$; in general, the answer is no. For example, if $F = \mathbb{Q}$ and $E = \mathbb{Q}(\alpha)$, where α is the real cube root of 2, then $G = \mathrm{Gal}(E/F) = \mathrm{Gal}(\mathbb{Q}(\alpha)/\mathbb{Q}) = \{1\}$ (if $\sigma \in G$, then $\sigma(\alpha)$ is a root of $x^3 - 2$; but E does not contain the other two (complex) roots of this polynomial). Hence $E^G = E \neq F$.

Theorem 60. *The following conditions are equivalent for a finite extension E/F with Galois group $G = \mathrm{Gal}(E/F)$.*

(i) $F = E^G$;

(ii) *every irreducible $p(x) \in F[x]$ with one root in E is separable and has all its roots in E; that is, $p(x)$ splits over E;*

(iii) *E is a splitting field of some separable polynomial $f(x) \in F[x]$.*

Proof. (i) $\Rightarrow$ (ii) Let $p(x) \in F[x]$ be an irreducible polynomial having a root α in E, and let the distinct elements of the set $\{\sigma(\alpha) : \sigma \in G\}$ be $\alpha_1, \ldots, \alpha_n$. Define $g(x) \in E[x]$ by

$$g(x) = \prod (x - \alpha_i).$$

Now each $\sigma \in G$ permutes the α_i; so that each σ fixes each of the coefficients of $g(x)$; that is, the coefficients of $g(x)$ lie in $E^G = F$. Hence $g(x)$ is a polynomial in $F[x]$ having no repeated roots. Now $p(x)$ and $g(x)$ have a common root in E, and so their gcd in $E[x]$ is not 1; it follows from Corollary 8 that their gcd is not 1 in $F[x]$. Since $p(x)$ is irreducible, it must divide $g(x)$. Therefore $p(x)$ has no repeated roots, hence is separable, and it splits over E.

(ii) $\Rightarrow$ (iii) Choose $\alpha_1 \in E$ with $\alpha_1 \notin F$. Since E/F is a finite extension, α_1 must be algebraic over F; let $p_1(x) \in F[x]$ be its irreducible polynomial. By hypothesis, $p_1(x)$ is a separable polynomial which splits over E; let $K_1 \subset E$ be its splitting field. If $K_1 = E$, we are done. Otherwise, choose $\alpha_2 \in E$ with $\alpha_2 \notin K_1$. By hypothesis, there is a separable irreducible $p_2(x) \in F[x]$ having α_2 as a root. Let $K_2 \subset E$ be the splitting field of $p_1(x)p_2(x)$, a separable polynomial. If $K_2 = E$, we are done; otherwise, iterate this construction. This process must end with $K_m = E$ for some m because E/F is finite.

(iii) $\Rightarrow$ (i) By Theorem 37, $|G| = [E:F]$. But Theorem 58 gives $|G| = [E:E^G]$, so that $[E:F] = [E:E^G]$. Since $F \subset E^G$, it follows that $F = E^G$. $\square$

Definition. A finite field extension E/F is **Galois** (or *normal*) if it satisfies any of the equivalent conditions in Theorem 60.

Remark. Terminology is not yet standard; some authors call a Galois extension normal; others call an extension normal if it is the splitting field of some, not necessarily separable, polynomial.

Exercises

*85. If E/F is a Galois extension and B is an intermediate field, then E/B is a Galois extension.

86. If F has characteristic $\neq 2$ and E/F is a field extension with $[E:F] = 2$, then E/F is Galois.

87. Show that being Galois need not be transitive; that is, if $F \subset B \subset E$ and E/B and B/F are Galois, then E/F need not be Galois. (Hint: Consider $\mathbb{Q} \subset \mathbb{Q}(\alpha) \subset \mathbb{Q}(\beta)$, where α is a square root of 2 and β is a fourth root of 2.)

*88. Let $E = F(x_1, \ldots, x_n)$ and let S be the subfield of all symmetric functions. Prove that $[E:S] = n!$ and $\mathrm{Gal}(E/S) \cong S_n$. (Hint: Show that E/S is a splitting field of the separable polynomial $f(t) = \prod(t - x_i)$.)

89. Let E/F be a Galois extension and let $p(x) \in F[x]$ be irreducible. Show that all the irreducible factors of $p(x)$ in $E[x]$ have the same degree. (Hint: Use Exercise 81.)

*90. Given a field F and a finite group G of order n, show that there is a subfield $K \subset E = F(x_1, \ldots, x_n)$ with $\mathrm{Gal}(E/K) \cong G$. (Hint: Use Exercise 88 and Cayley's theorem (Theorem A24).)

There is a version of Galois extension for intermediate fields.

Definition. Let E/F be a Galois extension and let B, C be intermediate fields. If there exists an isomorphism $B \to C$ fixing F, then C is called a **conjugate** of B.

Lemma 61. *Let E/F be a Galois extension, and let B be an intermediate field. The following conditions are equivalent.*

(i) *B has no conjugates (other than B itself);*

(ii) *If $\sigma \in \mathrm{Gal}(E/F)$, then $\sigma|B \in \mathrm{Gal}(B/F)$;*

(iii) *B/F is a Galois extension.*

Proof. (i) $\Rightarrow$ (ii) Obvious.

(ii) $\Rightarrow$ (iii) Let $p(x) \in F[x]$ be an irreducible polynomial having a root β in B. Since $B \subset E$ and E/F is Galois, all the roots of $p(x)$ lie in E. Suppose there is a root $\beta' \in E$ with $\beta' \notin B$. By Theorem 32, there exists an isomorphism $\tau: F(\beta) \to F(\beta')$ fixing F which extends to $\sigma \in \mathrm{Gal}(E/F)$, because E/F is Galois. But $\sigma(B) \cong B$ and $\sigma(B) \neq B$, because $\beta' \in \sigma(B)$ and $\beta' \notin B$.

(iii) $\Rightarrow$ (i) B/F is a splitting field of some polynomial $f(x)$ over F, so that $B = F(\alpha_1, \ldots, \alpha_n)$, where $\alpha_1, \ldots, \alpha_n$ are all the roots of $f(x)$. Since every $\sigma \in \mathrm{Gal}(E/F)$ must carry a root of $f(x)$ into a root of $f(x)$, it follows that σ must send B to itself. $\square$

The Fundamental Theorem of Galois Theory

Definition. A **lattice** is a partially ordered set $(L, \leq)$ in which each pair of elements a, b of L has a least upper bound $a \vee b$ and a greatest lower bound $a \wedge b$.

Recall that L is a *partially ordered set* if $\leq$ is a reflexive, transitive, and antisymmetric binary relation. An element c is an *upper bound* of a, b if $a \leq c$ and $b \leq c$; an element d is a *least upper bound* of a, b if it is an upper bound with $d \leq c$ for every upper bound c. Greatest lower bound is defined analogously, reversing the inequalities.

Examples

1. If X is a set, let L be the family of all the subsets of X, and define $A \leq B$ to mean $A \subset B$. Then L is a lattice with

$$A \vee B = A \cup B \quad \text{and} \quad A \wedge B = A \cap B.$$

2. If G is a group, let $\text{Sub}(G)$ be the family of all the subgroups of G, and define $H \leq K$ to mean $H \subset K$. Then $\text{Sub}(G)$ is a lattice with $H \vee K$ the subgroup generated by H and K, and $H \wedge K = H \cap K$.

3. Let E/F be a field extension, let $\text{Lat}(E/F)$ be the family of all intermediate fields, and define $B \leq C$ to mean $B \subset C$. Then $\text{Lat}(E/F)$ is a lattice with $B \vee C$ their compositum and $B \wedge C = B \cap C$.

4. Let L be the set of all integers $n \geq 1$, and define $n \leq m$ to mean $n \mid m$. Then L is a lattice with $n \vee m$ and $n \wedge m$ the lcm and gcd of n and m, respectively.

Lemma 62. *If L and L' are lattices and $\gamma: L \to L'$ is an **order reversing** bijection ($a \leq b$ implies $\gamma(b) \leq \gamma(a)$), then*

$$\gamma(a \vee b) = \gamma(a) \wedge \gamma(b) \quad and \quad \gamma(a \wedge b) = \gamma(a) \vee \gamma(b).$$

Proof. Now $a, b \leq a \vee b$ implies $\gamma(a), \gamma(b) \geq \gamma(a \vee b)$; that is, $\gamma(a \vee b)$ is a lower bound of $\gamma(a), \gamma(b)$. It follows that $\gamma(a) \wedge \gamma(b) \geq \gamma(a \vee b)$; since γ is surjective, there is $c \in L$ with $\gamma(a) \wedge \gamma(b) = \gamma(c)$. Apply γ^{-1} (which is easily seen to be order reversing also) to obtain $a, b \leq c \leq a \vee b$. Hence $c = a \vee b$ and $\gamma(a \vee b) = \gamma(c) = \gamma(a) \wedge \gamma(b)$. A similar argument proves the other half of the statement. $\square$

Theorem 63 (Fundamental Theorem of Galois Theory). *Let E/F be a Galois extension with Galois group $G = \text{Gal}(E/F)$.*

(i) *The function $\gamma: \text{Sub}(G) \to \text{Lat}(E/F)$, defined by $H \mapsto E^H$, is an order reversing bijection with inverse $\delta: B \mapsto \text{Gal}(E/B)$.*

(ii) $E^{\text{Gal}(E/B)} = B$ *and* $\text{Gal}(E/E^H) = H$.

(iii) $E^{H \vee K} = E^H \cap E^K$ *and* $E^{H \cap K} = E^H \vee E^K$;

$$\text{Gal}(E/B \vee C) = \text{Gal}(E/G) \cap \text{Gal}(E/C) \quad and$$
$$\text{Gal}(E/B \cap C) = \text{Gal}(E/B) \vee \text{Gal}(E/C).$$

(iv) $[B: F] = [G: \text{Gal}(E/B)]$ *and* $[G: H] = [E^H: F]$.

(v) B/F *is a Galois extension if and only if $\text{Gal}(E/B)$ is a normal subgroup of G.*

Proof. (i) It is easy to see that γ is order reversing: $K \leq H$ implies $E^H \leq E^K$. That γ is injective is precisely the statement of Corollary 59. To see that γ is surjective, consider the composite

$$\text{Lat}(E/F) \xrightarrow{\delta} \text{Sub}(G) \xrightarrow{\gamma} \text{Lat}(E/F),$$

where δ is the map $B \mapsto \text{Gal}(E/B)$. Then $\gamma\delta: B \mapsto \text{Gal}(E/B) \mapsto E^{\text{Gal}(E/B)}$. By Exercise 85, E/F Galois implies that E/B is Galois for every intermediate field B; hence Theorem 60 gives $B = E^{\text{Gal}(E/B)}$; hence $\gamma\delta$ is the identity and γ is a surjection. It follows that γ is a bijection with inverse δ.

(ii) This is just the statement that $\delta\gamma$ and $\gamma\delta$ are identity functions.

(iii) The first pair of equations follows from Lemma 62 because γ is an order reversing bijection; the second pair follows because $\delta = \gamma^{-1}$ is also an order reversing bijection.

(iv) $[B:F] = [E:F]/[E:B] = |G|/|\text{Gal}(E/B)| = [G:\text{Gal}(E/B)]$,

so that the degree of B/F is the index of $\text{Gal}(E/B)$ in G. The second equation follows from setting $B = E^H$, because $\text{Gal}(E/E^H) = H$.

(v) If B/F is Galois, then we have seen, in Theorem 39, that $\text{Gal}(E/B)$ is a normal subgroup of G. Conversely, suppose that H is a normal subgroup of G; is E^H/F a Galois extension? If $\sigma \in G$, then $\tau\sigma(\alpha) = \sigma\tau'(\alpha)$ for some $\tau' \in H$, by normality of H in G, and $\sigma\tau'(\alpha) = \sigma(\alpha)$ because τ' fixes α. Therefore $\alpha \in E^H$ implies $\sigma(\alpha) \in E^H$; that is, $\sigma(E^H) \subset E^H$; indeed, $\sigma(E^H) = E^H$ because both have the same dimension over F. By Lemma 61, E^H/F is a Galois extension. $\square$

Applications

Corollary 64. *A Galois extension E/F has only finitely many intermediate fields.*

Proof. Its Galois group is finite, hence has only finitely many subgroups. $\square$

Theorem 65 (Steinitz). *A finite extension E/F is simple if and only if it has only finitely many intermediate fields.*

Proof. Assume that $E = F(\alpha)$ and let $p(x)$ be the irreducible polynomial of α over F. Given an intermediate field B, let $g(x)$ be the irreducible polynomial of α over B. If B' is the subfield of B generated by F and the coefficients of $g(x)$, then $g(x)$ is also irreducible over B'. Since $E = B(\alpha) = B'(\alpha)$, it follows that $[E:B] = [B(\alpha):B]$ and $[E:B'] = [B'(\alpha):B']$; hence $[E:B] = [E:B']$, for both equal the degree of $g(x)$. Therefore, $B = B'$ and B is completely determined by $g(x)$. But $g(x)$ is a divisor of $p(x)$; as there are only finitely many monic divisors of $p(x)$ over E, there are only finitely many intermediate fields B.

Assume that E/F has only finitely many intermediate fields. If F is finite, then Corollary 46 shows that E/F is simple: just adjoin a primitive element. We may, therefore, assume that F is infinite. Now $E = F(\alpha_1, \ldots, \alpha_n)$; by induction on n, it suffices to prove that $E = F(\alpha, \beta)$ is a simple extension.

Consider all elements γ of the form $\gamma = \alpha + a\beta$, where $a \in F$; there are infinitely many such γ because F is infinite. Since there are only finitely many intermediate fields, there are only finitely many fields of the form $F(\gamma)$. There thus exist distinct elements $a, a' \in F$ with $F(\gamma) = F(\gamma')$, where $\gamma' = \alpha + a'\beta$. Clearly, $F(\gamma) \subset F(\alpha, \beta)$. For the reverse inclusion, $F(\gamma) = F(\gamma')$ contains $\gamma - \gamma' = (a - a')\beta$. Since $a \neq a'$, we have $\beta \in F(\gamma)$. But now $\alpha = \gamma - a\beta \in F(\gamma)$, so that $F(\alpha, \beta) \subset F(\gamma)$, as desired. $\square$

Corollary 66. *If E/F is a finite simple extension and B is an intermediate field, then B/F is simple.*

Corollary 67 (Theorem of the Primitive Element). *Every Galois extension E/F is simple.*

Proof. Immediate from Corollary 64 and Theorem 65. $\square$

Using the proof of Theorem 65, it is easy to show that one may choose a primitive element of $F(\alpha_1, \ldots, \alpha_n)$ of the form $a_1\alpha_1 + \cdots + a_n\alpha_n$ for $a_i \in F$.

Corollary 68. *The Galois field $GF(p^n)$ has exactly one subfield of order p^d for every divisor d of n.*

Proof. We have seen in Theorem 48 that $\mathrm{Gal}(GF(p^n)/GF(p)) \cong \mathbb{Z}_n$; moreover, Lemma 41 shows that a cyclic group of order n has exactly one subgroup of order d for every divisor d of n. Now a subgroup of order d has index n/d, and so the Fundamental Theorem says that the corresponding intermediate field has degree n/d. But the numbers n/d, as d varies over all the divisors of n, themselves vary over all the divisors of n. $\square$

Even more is true. The lattice of all intermediate fields is the same as the lattice of all subgroups of $\mathbb{Z}_n$, and this, in turn, is the same as the lattice of all the divisors of n under lcm and gcd (a sublattice of the lattice of Example 4).

Corollary 69. *If E/F is a Galois extension whose Galois group $\mathrm{Gal}(E/F)$ is abelian, then every intermediate field B/F is a Galois extension.*

Proof. Every subgroup of an abelian group is a normal subgroup. $\square$

Corollary 70. *Let $f(x) \in F[x]$ be a separable polynomial, and let E/F be a splitting field. Let $f(x) = g(x)h(x)$ in $F[x]$, and let B/F and C/F be splitting fields of $g(x)$, $h(x)$, respectively, contained in E. If $B \cap C = F$ (such fields are called **linearly disjoint** over F), then*

$$\mathrm{Gal}(E/F) \cong \mathrm{Gal}(B/F) \times \mathrm{Gal}(C/F).$$

Proof. Recall that if H and K are subgroups of a group G, then G is their *direct product*, denoted by $G = H \times K$, if both H and K are normal, $H \cap K =$

$\{1\}$, and $H \vee K = HK = G$. Let $G = \mathrm{Gal}(E/F)$. Since B/F and C/F are Galois extensions, both $\mathrm{Gal}(E/B)$ and $\mathrm{Gal}(E/C)$ are normal subgroups of G. The hypothesis gives $B \vee C = E$, so that $\mathrm{Gal}(E/B) \cap \mathrm{Gal}(E/C) = \mathrm{Gal}(E/B \vee C) = \mathrm{Gal}(E/E) = \{1\}$. Also, $\mathrm{Gal}(E/B)\mathrm{Gal}(E/C) = \mathrm{Gal}(E/B \cap C) = \mathrm{Gal}(E/F) = G$. Hence G is a direct product. Finally, since $(H \times K)/H \cong K$, Theorem 39 gives $\mathrm{Gal}(E/C) \cong G/\mathrm{Gal}(E/B) \cong \mathrm{Gal}(B/F)$ and $\mathrm{Gal}(E/B) \cong \mathrm{Gal}(C/F)$. $\square$

The fundamental theorem can also suggest counterexamples, for it translates problems about fields (which are usually infinite structures) into problems about finite groups. For example, let E/F be a Galois extension, and let B, C be intermediate fields of degree 2^b and 2^c, respectively; is the degree of their compositum $B \vee C$ also a power of 2? If $G = \mathrm{Gal}(E/F)$ and H and K are the subgroups corresponding to B and C, respectively, then the fundamental theorem gives

$$[B \vee C : F] = [G : H \cap K].$$

The translated question is: If $[G : H] = 2^b$ and $[G : K] = 2^c$, then is $[G : H \cap K]$ a power of 2? In Exercise 88, we saw that there is a Galois extension E/F with $\mathrm{Gal}(E/F) \cong S_4$. Let H be the subgroup of all permutations of $\{1, 2, 3\}$ (that is, all $\sigma \in S_4$ with $\sigma(4) = 4$) and let K be the subgroup of all permutations of $\{2, 3, 4\}$. Now $[S_4 : H] = 4 = [S_4 : K]$, but $[S_4 : H \cap K] = 12$ (because $H \cap K = \{(1), (23)\}$ has order 2).

Exercise

91. (i) Let E/F be a Galois extension. Must there be an intermediate field of prime degree over F? (Hint: The alternating group A_6 has no subgroups of prime index (see Theorem A35).)

   (ii) Same question as in (i) with the added hypothesis that $\mathrm{Gal}(E/F)$ is a solvable group.

We are now going to prove the fundamental theorem of algebra. Assume that $\mathbb{R}$ satisfies a weak form of the intermediate value theorem: if $f(x) \in \mathbb{R}[x]$ and there exist $a, b \in \mathbb{R}$ such that $f(a) > 0$ and $f(b) < 0$, then $f(x)$ has a real root. Here are some preliminary consequences.

(1) Every positive real r has a real square root.

If $f(x) = x^2 - r$, then $f(1 + r) > 0$ and $f(0) < 0$.

(2) Every quadratic $g(x) \in \mathbb{C}[x]$ has a complex root.

First, every complex number has a complex square root: one must find reals x, y with $(x + iy)^2 = a + ib$, and this is routine: the quadratic formula will be needed, and the expression under the radical (the discriminant) turns out to be a non-negative real. Another application of the quadratic formula now gives complex roots of $g(x)$.

(3) The field $\mathbb{C}$ has no extensions of degree 2.

Such an extension would contain an element whose irreducible polynomial is a quadratic in $\mathbb{C}[x]$, and (2) shows that no such polynomial exists.

(4) Every $f(x) \in \mathbb{R}[x]$ having odd degree has a real root.

Let $f(x) = a_0 + a_1 x + \ldots + a_{n-1} x^{n-1} + x^n \in \mathbb{R}[x]$. Define $t = 1 + \sum |a_i|$. Now $|a_i| \leq t - 1$ for all i, and

$$|a_0 + a_1 t + \ldots a_{n-1} t^{n-1}| \leq (t-1)[1 + t + \ldots + t^{n-1}]$$
$$= t^n - 1$$
$$< t^n.$$

It follows that $f(t) > 0$ (for any not necessarily odd n) because the sum of the early terms is dominated by t^n. When n is odd, $f(-t) < 0$, for

$$(-t)^n = (-1)^n t^n < 0,$$

and so the same estimate as above now shows that $f(-t) < 0$.

(5) There is no field extension $E/\mathbb{R}$ of odd degree > 1.

If $\alpha \in E$, then its irreducible polynomial must have even degree, by (4), so that $[\mathbb{R}(\alpha):\mathbb{R}]$ is even. Hence $[E:\mathbb{R}] = [E:\mathbb{R}(\alpha)][\mathbb{R}(\alpha):\mathbb{R}]$ is even.

Theorem 71 (Fundamental Theorem of Algebra). *Every nonconstant* $f(x) \in \mathbb{C}[x]$ *has a complex root.*

Proof. If $f(x) \in \mathbb{C}[x]$, then $f(x)\overline{f}(x) \in \mathbb{R}[x]$, where $\overline{f}(x)$ is obtained from $f(x)$ by taking the complex conjugate of every coefficient. Since $f(x)$ has a complex root if and only if $f(x)\overline{f}(x)$ has a complex root, it suffices to prove that every real polynomial has a complex root.

Let $p(x)$ be an irreducible polynomial in $\mathbb{R}[x]$, and let $E/\mathbb{R}$ be a splitting field of $(x^2 + 1)p(x)$ which contains $\mathbb{C}$. Since $\mathbb{R}$ has characteristic 0, $E/\mathbb{R}$ is a Galois extension; let G be its Galois group. If $|G| = 2^m k$, where k is odd, then G has a subgroup H of order 2^m, by the Sylow theorem (Theorem A13); let $B = E^H$ be the corresponding intermediate field. Now the degree $[B:\mathbb{R}]$ equals the index $[G:H] = k$. But we have seen above that $\mathbb{R}$ has no extension of odd degree > 1; hence $k = 1$ and G is a 2-group. By Theorem A23, the subgroup $\text{Gal}(E/\mathbb{C})$ of G (corresponding to $\mathbb{C}$) has a subgroup of index 2; its corresponding intermediate field is an extension of $\mathbb{C}$ of degree 2, and this contradicts (3) above. We conclude that $\text{Gal}(E/\mathbb{C}) = \{1\}$ and $E = \mathbb{C}$. $\square$

Galois's Great Theorem

We prove the converse of Theorem 53 (which holds only in characteristic 0): solvability of the Galois group implies solvability by radicals of the polynomial. We begin with some lemmas; the first one has a quaint name signifying its use as a device to get around the possible absence of roots of unity in the ground field.

Lemma 72 (Accessory Irrationalities). *Let E/F be a splitting field of $f(x) \in F[x]$ with Galois group $G = \mathrm{Gal}(E/F)$. If F^*/F is an extension and E^*/F^* is a splitting field of $f(x)$ containing E, then restriction $\sigma \mapsto \sigma|E$ is an injective homomorphism*

$$\mathrm{Gal}(E^*/F^*) \to \mathrm{Gal}(E/F).$$

Proof. The hypothesis gives $E = F(\alpha_1, \ldots, \alpha_n)$ and $E^* = F^*(\alpha_1, \ldots, \alpha_n)$, where $\{\alpha_1, \ldots, \alpha_n\}$ are the roots of $f(x)$. If $\sigma \in \mathrm{Gal}(E^*/F^*)$, then σ permutes $\{\alpha_1, \ldots, \alpha_n\}$ and fixes F^*, hence F; therefore, $\sigma|E \in \mathrm{Gal}(E/F)$. Using Exercise 73, one sees that $\sigma \mapsto \sigma|E$ is an injection. $\square$

Definition. If E/F is a Galois extension and $\alpha \in E^{\#}$, define its **norm** $N(\alpha)$ by

$$N(\alpha) = \prod_{\sigma \in G} \sigma(\alpha).$$

Here are some preliminary properties of the norm whose simple proofs are left as exercises.

(i) If $\alpha \in E^{\#}$, then $N(\alpha) \in F^{\#}$ (because $N(\alpha) \in E^G = F$).

(ii) $N(\alpha\beta) = N(\alpha)N(\beta)$, so that $N: E^{\#} \to F^{\#}$ is a homomorphism.

(iii) If $a \in F$, then $N(a) = a^n$, where $n = [E:F]$.

(iv) If $\sigma \in G$ and $\alpha \in E^{\#}$, then $N(\sigma(\alpha)) = N(\alpha)$.

Given a homomorphism, one asks about its kernel and image. The image of the norm is not easy to compute; the next result (which was the ninetieth theorem is an exposition of Hilbert (1897) on algebraic number theory) computes the kernel of the norm in a special case.

Lemma 73 (Hilbert's Theorem 90). *Let E/F be a Galois extension whose Galois group $G = \mathrm{Gal}(E/F)$ is cyclic of order n; let σ be a generator of G. Then $N(\alpha) = 1$ if and only if there exists $\beta \in E$ with*

$$\alpha = \beta\sigma(\beta)^{-1}.$$

Proof. If $\alpha = \beta\sigma(\beta)^{-1}$, then

$$N(\alpha) = N(\beta\sigma(\beta)^{-1}) = N(\beta)N(\sigma(\beta)^{-1})$$
$$= N(\beta)N(\sigma(\beta))^{-1} = N(\beta)N(\beta)^{-1} = 1.$$

For the converse, define "partial norms":

$$\delta_0 = \alpha, \quad \delta_1 = \alpha\sigma(\alpha), \quad \delta_2 = \alpha\sigma(\alpha)\sigma^2(\alpha), \ldots ,$$
$$\delta_{n-1} = \alpha\sigma(\alpha)\cdots\sigma^{n-1}(\alpha) = N(\alpha) = 1.$$

It is easy to see that

(1) $$\alpha\sigma(\delta_i) = \delta_{i+1} \quad \text{for all} \quad 0 \le i \le n-2.$$

By independence of the characters $\{1, \sigma, \sigma^2, \ldots, \sigma^{n-1}\}$, there exists $\gamma \in E$ with

$$\delta_0\gamma + \delta_1\sigma(\gamma) + \cdots + \delta_i\sigma^i(\gamma) + \cdots + \delta_{n-2}\sigma^{n-2}(\gamma) + \sigma^{n-1}(\gamma) \ne 0;$$

call this sum β. Using Eq. (1), one easily checks that

$$\sigma(\beta) = \alpha^{-1}[\delta_1\sigma(\gamma) + \cdots + \delta_i\sigma^i(\gamma) + \cdots + \delta_{n-1}\sigma^{n-1}(\gamma)] + \sigma^n(\gamma).$$

But $\sigma^n = 1$, so that the last term is just $\gamma = \alpha^{-1}\delta_0\gamma$. Hence $\sigma(\beta) = \alpha^{-1}\beta$, as desired. $\square$

Corollary 74. *Let E/F be a Galois extension of prime degree p. If F has a primitive pth root of unity, then $E = F(\beta)$, where $\beta^p \in F$, and so E/F is a pure extension.*

Proof. If ω is a primitive pth root of unity, then $N(\omega) = \omega^p = 1$, because $\omega \in F$. Now $G = \text{Gal}(E/F)$ has order p, hence is cyclic; let σ be a generator. By Hilbert's Theorem 90, we have $\omega = \beta\sigma(\beta)^{-1}$ for some $\beta \in E$. Hence $\sigma(\beta) = \beta\omega^{-1}$. It follows that $\sigma(\beta^p) = (\beta\omega^{-1})^p = \beta^p$, and so $\beta^p \in E^G = F$ because σ generates G and E/F is Galois. Note that $\beta \notin F$, lest $\omega = 1$, so that $F(\beta) \ne F$ is an intermediate field. Therefore $E = F(\beta)$, because $[E:F] = p$, and hence E has no proper intermediate fields. $\square$

Here is the converse of Theorem 53.

Theorem 75 (Galois). *Let F be a field of characteristic 0, let E/F be a Galois extension, and let $G = \text{Gal}(E/F)$ be a solvable group. Then E can be imbedded in a radical extension of F.*

Therefore, the Galois group of a polynomial over a field of characteristic 0 is a solvable group if and only if the polynomial is solvable by radicals.

Proof. The proof is by induction on $[E:F]$. The base step is trivially true. Since G is solvable, Corollary A17 provides a normal subgroup H of prime index, say, p. Let ω be a primitive pth root of unity (which exists because F has characteristic 0), and define $F^* = F(\omega)$ and $E^* = E(\omega)$. Observe that E^*/F is a Galois extension (if E/F is a splitting field of $f(x) \in F[x]$, then E^*/F is a splitting field of the necessarily separable polynomial $f(x)(x^p - 1)$); hence E^*/F^* is also a Galois extension. If E^* can be imbedded in a radical extension R^*/F^*, then E can also be imbedded in R^*. But F^*/F is a pure extension, so that R^* is a radical extension of F; therefore E can be imbedded in a radical extension of F, as desired.

Let $G^* = \text{Gal}(E^*/F^*)$. By accessory irrationalities, there is an injection $\psi: G^* \to G = \text{Gal}(E/F)$ (namely $\sigma \mapsto \sigma|E$), and so G^* is solvable (it is isomorphic to a subgroup of a solvable group). Consider the subgroup $\text{im}\,\psi$ ($\cong G^*$) of G. If $\text{im}\,\psi$ is a proper subgroup, then $[E^*:F^*] = |G^*| < |G| = [E:F]$. By induction, E^* can be imbedded in a radical extension of F^*. If $\text{im}\,\psi = G$, let $B = E^K$ (where $K = \psi^{-1}(H) \subset G^*$); then B/F^* is a Galois extension of prime degree p and F^* is a pure extension. Now E^*/B is a Galois extension of degree less than $[E^*:F^*] = |G^*| = |G| = [E:F]$. Since $\text{Gal}(E^*/B)$ is solvable (it is a subgroup of the solvable group G^*), the inductive hypothesis shows that E^* can be imbedded in a radical extension R' of B. As B/F^* is a pure extension, we see that R'/F^* is also a radical extension. $\square$

An earlier theorem of Abel, superseded by Galois, is that a polynomial with a commutative Galois group is solvable by radicals (because of this theorem, such groups are nowadays called abelian).

A deep theorem of Feit and Thompson (1963) says that every group of odd order is solvable. It follows that if F is a field of characteristic 0 and $f(x) \in F(x)$ is a polynomial whose Galois group has odd order, equivalently, whose splitting field has odd degree over F, then $f(x)$ is solvable by radicals.

Suppose one knows the Galois group G of a polynomial $f(x) \in \mathbb{Q}[x]$ and that G is solvable. Can one, in practice, use this information to find the roots of $f(x)$? The answer is affirmative; we suggest the reader look at the books of [Edwards] and [Gaal] to see how this is done.

Discriminants

Let F be a field of characteristic 0, let $f(x) \in F[x]$ be a polynomial of degree n having splitting field E/F, and let $G = \text{Gal}(E/F)$. If $\alpha_1, \ldots, \alpha_n$ are the roots of $f(x)$ in E (with repeated roots, if any, occurring several times), define

$$\Delta = \prod_{i<j}(\alpha_i - \alpha_j).$$

The number Δ depends on the indexing of the roots; a new indexing may change the sign of Δ. Therefore $D = \Delta^2$ depends only on the set of roots.

Definition. The **discriminant** of a polynomial $f(x) \in F[x]$ is $D = \Delta^2$.

It is clear that $f(x)$ has repeated roots if and only if $D = 0$. Each $\sigma \in G$ permutes $\alpha_1, \ldots, \alpha_n$, so that $\sigma(\Delta) = \pm\Delta$; hence $\Delta^2 = D \in E^G = F$.

If $f(x) = x^2 + bx + c$, then the quadratic formula gives the roots of $f(x)$:

$$\alpha = \frac{1}{2}\left(-b + \sqrt{b^2 - 4c}\right) \quad \text{and} \quad \beta = \frac{1}{2}\left(-b - \sqrt{b^2 - 4c}\right).$$

It follows that

$$D = \Delta^2 = (\alpha - \beta)^2 = b^2 - 4c.$$

If $f(x)$ is a cubic with roots α, β, γ, then

$$D = \Delta^2 = (\alpha - \beta)^2(\alpha - \gamma)^2(\beta - \gamma)^2;$$

it is not obvious how to compute D from the coefficients of $f(x)$.

Definition. A polynomial $f(x) = x^n + a_{n-1}x^{n-1} + \cdots + a_0$ is **reduced** if $a_{n-1} = 0$. If $f(x)$ is a monic polynomial of degree n, its **corresponding reduced polynomial** $\tilde{f}(x)$ is that obtained from $f(x)$ by the change of variable $y = x - a_{n-1}/n$.

Theorem 76. (i) *A polynomial $f(x)$ and its corresponding reduced polynomial $\tilde{f}(x)$ have the same discriminant.*

(ii) *The discriminant of a reduced cubic $\tilde{f}(x) = x^3 + qx + r$ is*

$$D = -4q^3 - 27r^2.$$

Proof. (i) If the roots of $f(x)$ are $\alpha_1, \ldots, \alpha_n$, then the roots of $\tilde{f}(x)$ are $\beta_1, \ldots, \beta_n$, where $\beta_i = \alpha_i + a_{n-1}/n$. Therefore

$$\prod_{i<j}(\alpha_i - \alpha_j) = \prod_{i<j}(\beta_i - \beta_j),$$

and so the discriminants (which are the squares of these) are equal.

(ii) The cubic formula gives the following roots of $\tilde{f}(x)$:

$$\alpha_1 = y + z; \quad \alpha_2 = \omega y + \omega^2 z; \quad \alpha_3 = \omega^2 y + \omega z;$$

here, ω is a cube root of unity, $y = \left[\frac{1}{2}(-r + \sqrt{R})\right]^{1/3}$, $z = \left[\frac{1}{2}(-r - \sqrt{R})\right]^{1/3}$, and $R = r^2 + 4q^3/27$. Hence

$$\Delta = (\alpha_1 - \alpha_2)(\alpha_1 - \alpha_3)(\alpha_2 - \alpha_3)$$
$$= (y + z - \omega y - \omega^2 z)(y + z - \omega^2 y - \omega z)(\omega y + \omega^2 z - \omega^2 y - \omega z).$$

It is easy to see:

$$\alpha_1 - \alpha_2 = (1 - \omega)(y - \omega^2 z);$$
$$\alpha_1 - \alpha_3 = -\omega^2(1 - \omega)(y - \omega z);$$
$$\alpha_2 - \alpha_3 = \omega(1 - \omega)(y - z);$$

hence

$$\Delta = -(1 - \omega)^3 \omega^3 (y - z)(y - \omega z)(y - \omega^2 z).$$

Now $\omega^3 = 1$ implies that $-(1 - \omega)^3 \omega^3 = 3\sqrt{3}i$ (where $i^2 = -1$); moreover, the identity

$$\zeta^3 - 1 = (\zeta - 1)(\zeta - \omega)(\zeta - \omega^2)$$

with $\zeta = y/z$, gives

$$(y - z)(y - \omega z)(y - \omega^2 z) = y^3 - z^3 = \sqrt{R}.$$

Therefore, $\Delta = 3\sqrt{3}i\sqrt{R}$ and

$$D = \Delta^2 = -27R = -27r^2 - 4q^3. \qquad \square$$

Exercises

*92. A polynomial and its corresponding reduced polynomial have the same Galois group.

93. (i) If $f(x) = x^3 + ax^2 + bx + c$, then its corresponding reduced polynomial is $x^3 + qx + r$, where

$$q = b - a^3/3 \quad \text{and} \quad r = 2a^3/27 - ab/3 + c.$$

   (ii) Show that the discriminant of $f(x)$ is

$$D = a^2 b^2 - 4b^3 - 4a^3 c - 27c^2 + 18abc.$$

Remark. There is a connection between the discriminant and the alternating group A_n. If $\pi \in S_n$, let π act on $\Delta = \prod_{i<j}(\alpha_i - \alpha_j)$ by permuting the subscripts; hence $\pi(\Delta) = \pm\Delta$. Define $\theta: S_n \to \mathbb{Z}_2$ by $\theta(\pi) = \overline{0}$ if $\pi(\Delta) = \Delta$, and $\theta(\pi) = \overline{1}$ if $\pi(\Delta) = -\Delta$. It is easy to see that θ is a surjective homomorphism with kernel A_n, for the alternating group is the unique subgroup of S_n having index 2 (Theorem A28).

Galois Groups of Quadratics, Cubics, and Quartics

In this final section, we show how to compute Galois groups of polynomials of low degree over $\mathbb{Q}$. Recall that the Galois group of a polynomial of degree n is a subgroup of S_n (regarded as the group of all permutations of the roots; of course, some of these permutations may have nothing to do with field automorphisms!).

Lemma 77. *Let $f(x) \in F[x]$ have discriminant $D = \Delta^2$ and Galois group $G = \text{Gal}(E/F)$. If $H = G \cap A_n$, then $E^H = F(\Delta)$; moreover, $\sqrt{D} \in F$ if and only if G is a subgroup of A_n.*

Proof. Clearly $F(\Delta) \subset E^H$ and $[E^H : F] = [G : H] \leq 2$; it suffices to prove that $[F(\Delta) : F] = [G : H]$. If $[G : H] = 2$, then there exists $\sigma \in G$, $\sigma \notin H$, with $\sigma(\Delta) \neq \Delta$; hence $\Delta \notin E^H = F$ and $[F(\Delta) : F] = 2$. If $[G : H] = 1$, then $G = H$ and $F(\Delta) \subset E^H = E^G = F$; hence $[F(\Delta) : F] = 1$.

For the second statement, the Fundamental Theorem of Galois Theory says that $F(\Delta) = E^H = F$ if and only if $G = H$ (because $E^G = F$). Since $H = G \cap A_n$, this means that $G \subset A_n$. $\square$

If $f(x) \in \mathbb{Q}[x]$ is quadratic, then its Galois group has order either 1 or 2 (because the symmetric group S_2 has order 2). The Galois group has order 1 if $f(x)$ splits; it has order 2 if $f(x)$ does not split; that is, if $f(x)$ is irreducible.

If $f(x) \in \mathbb{Q}[x]$ is a cubic having a rational root, then its Galois group G is the same as that of its quadratic factor. Otherwise $f(x)$ is irreducible; since $|G|$ is now a multiple of 3 and $G \subset S_3$, it follows that either $G \cong A_3 \cong \mathbb{Z}_3$ or $G \cong S_3$.

Theorem 78. *Let $f(x) \in \mathbb{Q}[x]$ be an irreducible cubic with Galois group G and discriminant D.*

(i) *$f(x)$ has exactly one real root if and only if $D < 0$, in which case $G \cong S_3$.*

(ii) *$f(x)$ has three real roots if and only if $D > 0$. In this case, either $\sqrt{D} \in \mathbb{Q}$ and $G \cong \mathbb{Z}_3$ or $\sqrt{D} \notin \mathbb{Q}$ and $G \cong S_3$.*

Proof. Note that $D \neq 0$ because $\mathbb{Q}$, having characteristic 0, is perfect, hence irreducible polynomials have no repeated roots. If $f(x)$ has three real roots, then Δ is real and $D = \Delta^2 > 0$. Conversely assume $f(x)$ has one real root α and two complex roots: $\beta = u + iv$ and $\bar{\beta} = u - iv$. By Theorem 76(i), the discriminant does not change when we eliminate the quadratic term of $f(x)$, so we may assume that this has been done; hence, $\alpha + \beta + \bar{\beta} = 0$ and $\alpha = -2u$. Therefore,

$$\begin{aligned}
\Delta &= (\alpha - \beta)(\alpha - \bar{\beta})(\beta - \bar{\beta}) \\
&= (-2u - u - iv)(-2u - u + iv)(2iv) \\
&= (9u^2 + v^2)2iv,
\end{aligned}$$

and so $D = \Delta^2 = -4v^2(9u^2 + v^2)^2 < 0$.

Let $E/\mathbb{Q}$ be the splitting field of $f(x)$. If $f(x)$ has exactly one real root α, then $E \neq \mathbb{Q}(\alpha)$. Hence $|G| > 3$ and $G \cong S_3$. If $f(x)$ has three real roots, then $D > 0$ and $\sqrt{D}$ is real. By Lemma 77, $G \cong A_3 \cong \mathbb{Z}_3$ if and only if $\sqrt{D}$ is rational; hence $G \cong S_3$ if $\sqrt{D}$ is irrational. $\square$

Consider a (reduced) quartic $f(x) = x^4 + qx^2 + rx + s \in \mathbb{Q}[x]$; let $E/\mathbb{Q}$ be its splitting field and let $G = \mathrm{Gal}(E/\mathbb{Q})$ be its Galois group. (By Exercise 92, it is no loss in generality to assume $f(x)$ is reduced.) If $f(x)$ has a rational root r, then $f(x) = (x - r)h(x)$, and its Galois group is the same as that of its cubic factor $h(x)$; this is computed by Theorem 78. Suppose that $f(x) = p(x)q(x)$ is the product of two irreducible quadratics; let α be a root of $p(x)$ and let β be a root of $q(x)$. If $\mathbb{Q}(\alpha) \cap \mathbb{Q}(\beta) = \mathbb{Q}$, that is, if these fields are linearly disjoint, then Corollary 70 shows that $G \cong V$, the four group; otherwise, $\alpha \in \mathbb{Q}(\beta)$, so that $\mathbb{Q}(\beta) = \mathbb{Q}(\alpha, \beta) = E$, and G has order 2.

We are left with the case $f(x)$ irreducible. The basic idea now is to compare G with the four group, namely, the normal subgroup of S_4:

$$V = \{(1), (12)(34), (13)(24), (14)(23)\},$$

so that we can identify the fixed field of $V \cap G$. If the four (distinct) roots of $f(x)$ are α_1, α_2, α_3, α_4, then consider the numbers:

$$u = (\alpha_1 + \alpha_2)(\alpha_3 + \alpha_4),$$
$$v = (\alpha_1 + \alpha_3)(\alpha_2 + \alpha_4),$$
$$w = (\alpha_1 + \alpha_4)(\alpha_2 + \alpha_3).$$

It is clear that if $\sigma \in V \cap G$, then σ fixes u, v, and w. Conversely, checking each of the 24 permutations shows that if $\sigma \in S_4$ fixes $(\alpha_i + \alpha_j)(\alpha_k + \alpha_\ell)$, then $\sigma \in V \cup \{(ij), (k\ell), (ikj\ell), (i\ell jk)\}$. It follows that $\sigma \in G$ fixes each of u, v, w if and only if $\sigma \in V \cap G$, and so $\mathbb{Q}(u, v, w)$ is the fixed field of $V \cap G$.

Definition. The **resolvent cubic**[7] of $f(x) = x^4 + qx^2 + rx + s$ is

$$g(x) = (x - u)(x - v)(x - w).$$

---

[7]There is another resolvent cubic in the literature which arises from another combination of the roots invariant under V. Define

$$u' = \alpha_1\alpha_2 + \alpha_3\alpha_4,$$
$$v' = \alpha_1\alpha_3 + \alpha_2\alpha_4,$$
$$w' = \alpha_1\alpha_4 + \alpha_2\alpha_3,$$

and define $h(x) = (x - u')(x - v')(x - w')$. This cubic (which is distinct from $g(x)$ above) behaves much as $g(x)$ does. The reason for our preference for $g(x)$ is Exercise 94; one can use $g(x)$ to compute the discriminant of a quartic.

Theorem 79. *If $g(x)$ is the resolvent cubic of $f(x) = x^4 + qx^2 + rx + s$, then*

$$g(x) = x^3 - 2qx^2 + (q^2 - 4s)x + r^2.$$

Proof. In our discussion of the classical quartic formula, we saw that $f(x) = (x^2 + kx + \ell)(x^2 - kx + m)$ and k^2 is a root of

$$h(x) = x^3 + 2qx^2 + (q^2 - 4s)x - r^2,$$

a polynomial differing from the claimed expression for $g(x)$ only in the sign of its quadratic and constant terms. Thus, a number β is a root of $h(x)$ if and only if $-\beta$ is a root of $g(x)$.

Let the four roots $\alpha_1, \alpha_2, \alpha_3, \alpha_4$ of $f(x)$ be indexed so that α_1, α_2 are roots of $x^2 + kx + \ell$ and α_3, α_4 are roots of $x^2 - kx + m$. Then $k = -(\alpha_1 + \alpha_2)$ and $-k = -(\alpha_3 + \alpha_4)$; therefore

$$u = (\alpha_1 + \alpha_2)(\alpha_3 + \alpha_4) = -k^2$$

and $-u$ is a root of $h(x)$.

Now factor $f(x)$ into two quadratics, say,

$$f(x) = (x^2 + \tilde{k}x + \tilde{\ell})(x^2 - \tilde{k}x + \tilde{m}),$$

where α_1, α_3 are roots of the first factor and α_2, α_4 are roots of the second. The same argument as above now shows that

$$v = (\alpha_1 + \alpha_3)(\alpha_2 + \alpha_3) = -\tilde{k}^2,$$

hence $-v$ is a root of $h(x)$. Similarly, $-w = -(\alpha_1 + \alpha_4)(\alpha_2 + \alpha_3)$ is a root of $h(x)$. Therefore

$$h(x) = (x + u)(x + v)(x + w),$$

and so

$$g(x) = (x - u)(x - v)(x - w)$$

is obtained from $h(x)$ by changing the sign of the quadratic and constant terms. $\square$

Exercises

*94. If $f(x)$ is a quartic, then its discriminant is the negative of the discriminant of its resolvent cubic. (Hint:

$$u - v = -(\alpha_1 - \alpha_4)(\alpha_2 - \alpha_3)$$
$$u - w = -(\alpha_1 - \alpha_3)(\alpha_2 - \alpha_4)$$
$$v - w = -(\alpha_1 - \alpha_2)(\alpha_3 - \alpha_4).)$$

95. If the resolvent cubic of a quartic $f(x)$ is $x^3 + ax^2 + bx + c$, then the discriminant of $f(x)$ is

$$D = -16a^4c + 4a^3b^2 + 128a^2c^2 - 144ab^2c + 27b^4 - 256c^3.$$

96. Show that $x^3 + ax + 2 \in \mathbb{R}[x]$ has three real roots if and only if $a \leq -3$.

*97. Let G be a subgroup of S_4 with $|G|$ a multiple of 4; define

$$m = |G/G \cap V|,$$

where V is the four group.

   (i) Prove that m is a divisor of 6.

   (ii) If $m = 6$, then $G = S_4$; if $m = 3$, then $G = A_4$; if $m = 1$, then $G = V$; if $m = 2$, then $G \cong D_8$, $G \cong \mathbb{Z}_4$, or $G \cong V$.

(Hint: This exercise in group theory is Theorem A33.)

*98. Let G be a subgroup of S_4 with $|G| = 4$. If G acts transitively on $X = \{1, 2, 3, 4\}$ and $|G \cap V| = 2$, then $G \cong D_8$ or $G \cong \mathbb{Z}_4$. (**Remark.** If G acts transitively on X, then $|G|$ is a multiple of 4 (Lemma A10). This stronger hypothesis thus removes the possibility $G \cong V$ when $m = 2$ in Exercise 97.)

Theorem 80. *Let $f(x) \in \mathbb{Q}[x]$ be an irreducible quartic with Galois group G, and let m be the order of the Galois group of its resolvent cubic.*

   (i) *If $m = 6$, then $G \cong S_4$.*

   (ii) *If $m = 3$, then $G \cong A_4$.*

   (iii) *If $m = 1$, then $G \cong V$.*

   (iv) *If $m = 2$, then $G \cong D_8$ or $G \cong \mathbb{Z}_4$.*

Remark. Note that, in the ambiguous case (iv), the two possible groups have different orders. See Exercise 104.

Proof. We have seen that $\mathbb{Q}(u, v, w)$ is the fixed field of $V \cap G$. By the Fundamental Theorem,

$$|G/V \cap G| = [G : V \cap G] = [\mathbb{Q}(u, v, w) : \mathbb{Q}] = |\mathrm{Gal}(\mathbb{Q}(u, v, w)/\mathbb{Q})|.$$

Since $f(x)$ is irreducible, G acts transitively on its roots, by Exercise 84, hence $|G|$ is divisible by 4 (Lemma A10), and the theorem follows from Exercises 97 and 98. $\square$

Examples

1. Let $f(x) = x^4 - 4x + 2 \in \mathbb{Q}[x]$; $f(x)$ is irreducible, by Eisenstein's criterion. The resolvent cubic is $g(x) = x^3 - 8x + 16$. Now $g(x)$ is irreducible, for if one reduces mod 5, one obtains $x^3 + 2x + 1$, and this polynomial is irreducible over $\mathbb{Z}_5$ because it has no roots. The discriminant of $g(x)$ is -4864, so that Theorem 78 shows that the Galois group of $g(x)$ is S_3, hence has order 6. Theorem 80 now shows that $G \cong S_4$.

2. Let $f(x) = x^4 - 10x^2 + 1 \in \mathbb{Q}[x]$; $f(x)$ is irreducible, by Exercise 67. The resolvent cubic is $x^3 + 20x^2 + 96x = x(x + 8)(x + 12)$. In this case, $\mathbb{Q}(u, v, w) = \mathbb{Q}$ and $m = 1$. Therefore, $G \cong V$. (This should not be a surprise if one remembers that $f(x)$ arose as the irreducible polynomial of α, where $\mathbb{Q}(\alpha) = \mathbb{Q}(\sqrt{2}, \sqrt{3})$.)

Remark. If d is a divisor of $|S_4| = 24$, then it is known that S_4 has a subgroup of order d. If $d = 4$, then V and $\mathbb{Z}_4$ are nonisomorphic subgroups of order d; for any other divisor d, any two subgroups of order d are isomorphic. We conclude that the Galois group G of a quartic is determined to isomorphism by its order unless $|G| = 4$.

Exercises

99. Compute the Galois group over $\mathbb{Q}$ of $x^4 + x^2 - 6$.

100. Compute the Galois group over $\mathbb{Q}$ of $x^4 + x^2 + x + 1$.

101. Compute the Galois group over $\mathbb{Q}$ of $4x^4 + 12x + 9$.

102. (i) A quintic polynomial is solvable by radicals if and only if its Galois group has order ≤ 24.

   (ii) An irreducible quintic is solvable by radicals if and only if its Galois group has order ≤ 20.

   (Hint: A subgroup G of S_5 is solvable if and only if $|G| \leq 24$; see Theorem A38.)

The next three exercises are from [Kaplansky (1972), p. 53].

103. Let $f(x) \in \mathbb{Q}[x]$ be an irreducible quartic with Galois group G. If $f(x)$ has exactly two real roots, then either $G \cong S_4$ or $G \cong D_8$.

104. Let $x^4 + ax^2 + b$ be an irreducible polynomial over $\mathbb{Q}$ having Galois group G.

   (i) If b is a square in $\mathbb{Q}$, then $G \cong V$.

   (ii) If b is not a square in $\mathbb{Q}$ but $b(a^2 - 4b)$ is a square, then $G \cong \mathbb{Z}_4$.

(iii) If neither b nor $b(a^2 - 4b)$ is a square, then $G \cong D_8$.

105. Let $x^4 + bx^3 + cx^2 + bx + 1 \in \mathbb{Q}[x]$ have Galois group G.

   (i) If $h = c^2 + 4c + 4 - 4b^2$ is a square in $\mathbb{Q}$, then $G \cong V$.

   (ii) If h is not a square in $\mathbb{Q}$ but $h(b^2 - 4c + 8)$ is a square, then $G \cong \mathbb{Z}_4$.

   (iii) If neither h nor $h(b^2 - 4c + 8)$ is a square in $\mathbb{Q}$, then $G \cong D_8$.

106. If a herring and a half cost a penny and a half, how much does a dozen herring cost? (Answer: One shilling.)

Epilogue

You have seen an introduction to Galois theory; of course, there is more. A deeper study of *abelian fields,* that is, fields having (possibly infinite) abelian Galois groups, begins with *Kummer theory* and continues on to *class field theory.* Infinite Galois groups are topologized, and there is a bijection between intermediate fields and closed subgroups. The theorems are of basic importance in algebraic number theory. There is also a Galois theory classifying division algebras (see [Jacobson (1956)] and a Galois theory classifying commutative rings (see [Chase, Harrison, Rosenberg]).

An interesting open question is to determine which abstract finite groups G can be realized as Galois groups over $\mathbb{Q}$ (Exercise 90 shows that G can always be realized over some ground field). Many special examples have long been known. For example, the symmetric and alternating groups can be realized over $\mathbb{Q}$ (a proof for S_n can be found in [Hadlock, p. 210]); for a proof that the quarternions can be realized, see [R.A. Dean, Amer. Math. Monthly (1981), pp. 42–45] where it is shown to be the Galois group of $x^8 - 72x^6 + 180x^4 - 144x^2 + 36$). It is a deep result of Shafarevich (1954) that every solvable group can be realized. After the classification of the finite simple groups in the 1980's, there were attempts to realize them, with much success. However, it is still not known whether every finite simple group is a Galois group over $\mathbb{Q}$.

There is Galois theory in complex variables (see [Miller, Blichfeldt, Dickson, Chap. XX, p. 378]). In 1850, Puiseux studied the *monodromy group* of a certain class of functions $f(t, z)$ of two complex variables, namely, $f(t, z) \in \mathbb{C}(t)[z]$; in 1851, Hermite showed that this monodromy group is isomorphic to the Galois group of $f(t, z)$ over the function field $\mathbb{C}(t)$.

There is Galois theory in differential equations, due to Ritt and Kolchin (see [Kaplansky (1957)]). A *derivation* of a field F is an additive homomorphism $D: F \to F$ with $D(xy) = xD(y) + D(x)y$; an ordered pair (F, D) is called a *differential field.* Given a differential field (F, D) with F a (possibly infinite) extension of $\mathbb{C}$, its *differential Galois group* is the subgroup of

$\mathrm{Gal}(F/\mathbb{C})$ consisting of all σ commuting with D. If this group is suitably topologized and if the extension $F/\mathbb{C}$ satisfies conditions analogous to being a Galois extension (it is called a *Picard–Vessiot extension*), then there is a bijection between the intermediate differential fields and the closed subgroups of the differential Galois group.

There is Galois theory in algebraic topology. A *covering space* of a topological space X is an ordered pair $(\tilde{X}, p)$, where $p\colon \tilde{X} \to X$ is a certain type of continuous map. The elements of the group $\mathrm{Cov}(\tilde{X}/X)$ defined as {homeomorphisms $h\colon \tilde{X} \to \tilde{X} : ph = p$} are dual to the elements of a Galois group in the following sense. If $i\colon F \to E$ is the inclusion, where E/F is a Galois extension, then an automorphism σ of E lies in the Galois group if and only if $\sigma i = i$. When $\tilde{X}$ is simply connected, then $\mathrm{Cov}(\tilde{X}/X) \cong \pi_1(X)$, the fundamental group of X; moreover, there is a bijection between the family of all covering spaces of X and the family of all subgroups of the fundamental group.

I am awed by the genius of Galois (1811–1832). He solved one of the outstanding mathematical problems of his time, and his solution is beautiful; in so doing, he created two powerful theories, group theory and Galois theory, and his work is still influential today. And he did all of this at the age of 19; he was killed a year later.

Appendix 1
Group Theory Dictionary

Abelian group. A group in which multiplication is commutative.

Alternating group A_n. The subgroup of S_n consisting of all the even permutations. It has order $n!/2$.

Associativity. For all x, y, z, one has $(xy)z = x(yz)$. It follows that one does not need parentheses for any product of three or more factors.

Automorphism. An isomorphism of a group with itself.

Commutativity. For all x, y, one has $xy = yx$.

Coset of H in G. A subset of G of the form $gH = \{gh : h \in H\}$, where H is a subgroup of G and $g \in G$. All the cosets of H partition G; moreover, $gH = g'H$ if and only if $g^{-1}g' \in H$.

Cyclic group. A group G which contains an element g (called a generator) such that every element of G is some power of g.

Dihedral group D_{2n}. A group of order $2n$ containing an element a of order n and an element b of order 2 such that $bab = a^{-1}$.

Even permutation. A permutation that is a product of an even number of transpositions. Every r-cycle, for r odd, is an even permutation.

Factor groups of a normal series $G = G_0 \supset G_1 \supset \ldots \supset G_n = \{1\}$. The groups G_i/G_{i+1}.

Four group V. The subgroup of S_4 consisting of the identity, (12)(34), (13)(24), and (14)(23); it is a normal subgroup.

Generator of a cyclic group G. An element $g \in G$ whose powers give all the elements of G; a cyclic group may have several different generators.

Group. A set G equipped with an associative multiplication such that there is a unique $e \in G$ (called the *identity* of G) with $ex = x = xe$ for all $x \in G$, and, for each $x \in G$, there is a unique $y \in G$ (called the *inverse* of x) with $yx = e = xy$. One usually denotes e by 1 and y by x^{-1}. (Some of these axioms are redundant.)

Homomorphism. A function $f: G \to H$, where G and H are groups, such that $f(xy) = f(x)f(y)$ for all $x, y \in G$. One always has $f(1) = 1$ and $f(x^{-1}) = f(x)^{-1}$.

Image of a homomorphism $f: G \to H$. The subgroup of H consisting of all $f(x)$ for $x \in G$.

Index $[G:H]$. The number of cosets of a subgroup H in G; it is equal to $|G|/|H|$.

Isomorphism. A homomorphism that is a bijection.

Kernel of a homomorphism $f: G \to H$. The normal subgroup of G consisting of all $x \in G$ with $f(x) = 1$.

Natural map. If H is a normal subgroup of G, then the natural map is the homomorphism $\pi: G \to G/H$ defined by $\pi(x) = xH$.

Normal series of G. A sequence of subgroups $G = G_0 \supset G_1 \supset \ldots \supset G_n = \{1\}$ with each G_{i+1} a normal subgroup of G_i. (A subgroup G_i may not be a normal subgroup of G.)

Normal subgroup. A subgroup H of a group G such that $gHg^{-1} = H$ for all $g \in G$.

Order of an element $x \in G$. The least positive integer m, if any, such that $x^m = 1$; otherwise infinity.

Order $|G|$ of a group G. The number of elements in G.

p-group. A finite group of order a power of some prime p.

Permutation. A bijection of a set to itself; all the permutations form a group under composition.

Quotient group G/H. If H is a normal subgroup of G, it is the family of all cosets gH of H with multiplication defined by $gHg'H = gg'H$; the order of G/H is $|G|/|H|$; the identity element is $1H = H$; the inverse of gH is $g^{-1}H$.

Simple group G. A group $\neq \{1\}$ whose only normal subgroups are $\{1\}$ and G.

Solvable group. A group having a normal series with abelian factor groups.

Subgroup H *of* G. A subset of G containing 1 which is closed under multiplication and inverse.

Subgroup generated by a subset X. The smallest subgroup of G containing X; it consists of all products $x_1^a x_2^b \ldots x_n^z$, where $x_i \in X$ and the exponents $= \pm 1$.

Sylow p-subgroup of G. A subgroup of G of order p^n, where p^n is the highest power of p dividing $|G|$. Such subgroups always exist, and any two such are conjugate, hence isomorphic.

Symmetric group S_n. The group of all permutations of $\{1, 2, \ldots, n\}$ under composition; it has order $n!$.

Appendix 2
Group Theory Used in the Text

All groups in this appendix are assumed to be finite even though several of the theorems hold (perhaps with different proofs) in the infinite case as well. Definitions of terms can be found in the dictionary, Appendix 1.

Theorem A1. *Every subgroup S of a cyclic group $G = \langle a \rangle$ is itself cyclic.*

Proof. If $S = \{1\}$, then S is cyclic with generator 1. Otherwise, let m be the least positive integer for which $a^m \in S$; we claim $S = \langle a^m \rangle$. Clearly $\langle a^m \rangle \subset S$. For the reverse inclusion, take $s = a^k \in S$. By the division algorithm, there are integers q and r with $0 \le r < m$ and

$$k = qm + r.$$

But $a^k = a^{qm+r} = (a^m)^q a^r$ gives $a^r \in S$. If $r > 0$, the minimality of m is contradicted; therefore $r = 0$ and $a^k = (a^m)^q \in \langle a^m \rangle$. $\qquad \square$

Theorem A2. (i) *If $a \in G$ is an element of order n, then $a^m = 1$ if and only if $n \mid m$.*

(ii) *If $G = \langle a \rangle$ is a cyclic group of order n, then a^k is a generator of G if and only if $(k, n) = 1$.*

Proof. (i) The division algorithm provides integers q and r with $m = nq+r$, when $0 \le r < n$. It follows that $a^r = a^{m-nq} = a^m a^{-nq} = 1$. If $r > 0$, then we contradict n being the smallest positive integer with $a^n = 1$. Hence $r = 0$ and $n \mid m$.

(ii) Recall that two integers are relatively prime if and only if some integral linear combination of them is 1.

If a^k generates G, then $a \in \langle a^k \rangle$, so that $a = a^{kt}$ for some $t \in \mathbb{Z}$. Therefore $a^{kt-1} = 1$; by (i), $n \mid kt - 1$, so there is $v \in \mathbb{Z}$ with $nv = kt - 1$; that is, $(k, n) = 1$.

Conversely, if $(k, n) = 1$, then $nt + ku = 1$ for $t, u \in \mathbb{Z}$; hence $a = a^{nt+ku} = a^{nt}a^{ku} = a^{ku} \in \langle a^k \rangle$. Therefore every power of a also lies in $\langle a^k \rangle$ and $G = \langle a^k \rangle$.   □

Theorem A3 (Lagrange). *If H is a subgroup of a group G, then*

$$|G| = [G\colon H]|H|.$$

Proof. The cosets of H in G partition G (for the relation $x \sim y$, defined by $y = xh$ for some $h \in H$, is an equivalence relation on G whose equivalence classes are the cosets of H). Moreover $|H| = |xH|$ for every $x \in G$ (because $h \mapsto xh$ is a bijection), so that $|G|$ is the number of cosets times their common size.   □

It follows that $[G\colon H] = |G|/|H|$. In particular, if H is a normal subgroup of G (so that the quotient group G/H is defined), then

$$|G/H| = [G\colon H] = |G|/|H|.$$

Lemma A4. *Let $f\colon G \to H$ be a homomorphism with kernel K. Then f is an injection if and only if $K = \{1\}$.*

Proof. If f is an injection, then $x \neq 1$ implies $f(x) \neq f(1) = 1$, and so $x \notin K$. Conversely, assume $K = \{1\}$ and that $f(x) = f(y)$ for $x, y \in G$. Then

$$1 = f(x)f(y)^{-1} = f(x)f(y^{-1}) = f(xy^{-1})$$

and $xy^{-1} \in K = \{1\}$. Hence $x = y$ and f is an injection.   □

If $f\colon G \to H$ is a homomorphism, denote the image of f by $\operatorname{im} f$ and the kernel of f by $\ker f$.

Theorem A5 (First Isomorphism Theorem). *If $f\colon G \to H$ is a homomorphism, then $\ker f$ is a normal subgroup of G and*

$$G/\ker f \cong \operatorname{im} f.$$

Proof. Let $K = \ker f$. Let us show K is a subgroup. It does contain 1 (because $f(1) = 1$); if $x, y \in K$ (so that $f(x) = 1 = f(y)$), then $f(xy) = f(x)f(y) = 1$ and $xy \in K$; if $x \in K$, then $f(x^{-1}) = f(x)^{-1} = 1$ and $x^{-1} \in K$. Furthermore, the subgroup K is normal: if $x \in K$ and $g \in G$, then $f(gxg^{-1}) = f(g)f(x)f(g)^{-1} = f(g)f(g)^{-1} = 1$ and so $gxg^{-1} \in K$.

Define $\varphi\colon G/K \to \operatorname{im} f$ by $\varphi(xK) = f(x)$. Now φ is well defined: if $x'K = xK$, then $x' = xk$ for some $k \in K$, and $f(x') = f(xk) = f(x)f(k) = f(x)$. It is routine to check that φ is a homomorphism (because f is) with $\operatorname{im} \varphi = \operatorname{im} f$. Finally, φ is an injection, by Lemma A4, because $\varphi(xK) = 1$ implies $f(x) = 1$, hence $x \in K$ and $xK = K$.   □

If K, H are subgroups of G, then $K \vee H$ is the smallest subgroup of G containing K and H; that is, $K \vee H$ is the subgroup of G generated by $K \cup H$.

Lemma A6. *If K and H are subgroups of G with K normal in G, then $K \vee H = KH = \{kh: k \in K \text{ and } h \in H\} = HK$.*

Proof. Clearly $KH \subset K \vee H$. For the reverse inclusion, it suffices to prove that KH is a subgroup, for it does contain $K \cup H$.

Now $khk_1h_1 = k(hk_1h^{-1})hh_1 = (kk_2)(hh_1) \in KH$ for some $k_2 \in K$ (because K is normal). Also $(kh)^{-1} = h^{-1}k^{-1} = (h^{-1}k^{-1}h)h^{-1} = k'h^{-1} \in KH$ for some $k' \in K$ (again, because K is normal). Therefore, KH is a subgroup.

If $hk \in HK$, then $hk = (hkh^{-1})h = k'h \in KH$ for some $k' \in K$, and so $HK \subset KH$; the reverse inclusion is proved similarly. $\square$

If K and H are subgroups of G with K normal, then the family of those cosets hK of K with $h \in H$ is easily seen to be a subgroup of G/K. Indeed, one may check, using Lemma A6, that this subgroup is KH/K.

Theorem A7 (Second Isomorphism Theorem). *If K and H are subgroups of G with K normal in G, then $K \cap H$ is a normal subgroup of H and*

$$H/K \cap H \cong KH/K.$$

Proof. Let $\pi: G \to G/K$ be the natural map, defined by $\pi(x) = xK$, and let $f: H \to G/K$ be the restriction $\pi \mid H$. Now $\ker f = K \cap H$ and $\operatorname{im} f$ is the family of all cosets xK in G/K with $x \in H$ (hence $\operatorname{im} f = KH/K$). The first isomorphism theorem now gives the result. $\square$

Theorem A8 (Third Isomorphism Theorem). *If $S \subset K$ are normal subgroups of G, then K/S is a normal subgroup of G/S and*

$$(G/S)/(K/S) \cong G/K.$$

Proof. The function $f: G/S \to G/K$ given by $xS \mapsto xK$ is well defined because $S \subset K$. One checks easily that f is a surjective homomorphism with kernel K/S, and so the theorem follows from the first isomorphism theorem. $\square$

Theorem A9 (Correspondence Theorem). *Let K be a normal subgroup of G, and let S^* be a subgroup of $G^* = G/K$.*

(i) *There is a unique intermediate subgroup S, i.e., $K \subset S \subset G$, with $S/K = S^*$;*

(ii) *If S^* is a normal subgroup of G^*, then S is normal in G;*

(iii) $[G^*:S^*] = [G:S]$;

(iv) *If T^* is normal in S^*, then T is normal in S and*

$$S^*/T^* \cong S/T.$$

Proof. (i) Define $S = \{x \in G : xK \in S^*\}$.

(ii) If $a \in G$, and $x \in S$, then $axa^{-1}K = aKxKa^{-1}K \in S^*$, because S^* is normal in G^*; therefore $axa^{-1} \in S$.

(iii)

$$[G^*:S^*] = |G^*|/|S^*| = |G/K|/|S/K|$$
$$= |G|/|K|/|S|/|K| = |G|/|S| = [G:S].$$

(iv) T is normal in S, by (ii), and

$$S^*/T^* = (S/K)/(T/K) \cong S/T,$$

by the third isomorphism theorem. $\quad\square$

Definition. A group G **acts** on a set X if there is a function

$$G \times X \to X,$$

denoted by $(g, x) \mapsto g \cdot x$, such that:

(i) $1 \cdot x = x$ for all $x \in X$, where 1 is the identity in G;

(ii) $(gh) \cdot x = g \cdot (h \cdot x)$ for all $x \in X$, all $g, h \in G$.

Definition. If G acts on X and $x \in X$, then the **orbit** of x is

$$o(x) = \{gx : g \in G\} \subset X,$$

and the **stabilizer** of x is

$$G_x = \{g \in G : g \cdot x = x\} \subset G.$$

A group G acts **transitively** on X if, for each $x, y \in X$, there exists $g \in G$ with $g \cdot x = y$. In this case, $o(x) = X$.

Every group G acts on itself (here $X = G$) by conjugation: define $g \cdot x = gxg^{-1}$. The orbit $o(x)$ of $x \in G$ is its **conjugacy class**:

$$\{y \in G : y = gxg^{-1} \text{ for some } g \in G\};$$

the stabilizer of x is

$$\{g \in G : x = g \cdot x = gxg^{-1}\} = \{g \in G : gx = xg\}$$

(this last subgroup is called the **centralizer** of x in G; it is denoted by $C_G(x)$).

The reader may check that the family of all orbits partitions X, for the relation $x \sim y$ on X, defined by $y = g \cdot x$ for some $g \in G$, is an equivalence relation whose equivalence classes are the orbits.

Theorem A10. *If G acts on a set X with $|X| = n$ and if $x \in X$, then*

$$|o(x)| = [G{:}G_x].$$

In particular, if G acts transitively on X, then

$$|G| = n|G_x|.$$

Proof. Define $\varphi \colon o(x) \to \{$the family of all cosets of G_x in $G\}$ by

$$\varphi(g \cdot x) = gG_x.$$

Now φ is well defined, for if $g \cdot x = h \cdot x$ (where $g, h \in G$), then $h^{-1}g \cdot x = x$, $h^{-1}g \in G_x$, and $gG_x = hG_x$. Reversing this argument shows that φ is an injection: if $\varphi(g \cdot x) = \varphi(h \cdot x)$, then $gG_x = hG_x$, $h^{-1}g \in G_x$, and $g \cdot x = h \cdot x$. Finally, φ is surjective, for a coset gG_x is $\varphi(g \cdot x)$. Hence, φ is a bijection. If G acts transitively, then $o(x) = X$ and $|o(x)| = n = |X|$; hence $n = [G{:}G_x] = |G|/|G_x|$, and $|G| = n|G_x|$. $\quad\square$

Corollary A11. *If $x \in G$, then the number of its conjugates is $[G{:}C_G(x)]$.*

Proof. This is the special case of G acting on itself by conjugation. $\quad\square$

Lemma A12. *If p is a prime not dividing m ($p \nmid m$) and if $k \geq 1$, then*

$$p \nmid \binom{p^k m}{p^k}.$$

Proof. Write the binomial coefficient as follows:

$$\binom{p^k m}{p^k} = \frac{p^k m(p^k m - 1) \cdots (p^k m - i) \cdots (p^k m - p^k + 1)}{p^k(p^k - 1) \cdots (p^k - i) \cdots (p^k - p^k + 1)}.$$

Since p is prime, any factor p of the numerator (or of the denominator) arises from a factor of $p^k m - i$ (or of $p^k - i$). The highest power of p dividing $p^k m - i$, say, $p^{t(i)}$, is the same as the highest power of p dividing $p^k - i$ (because $p \nmid m$). Every factor of p upstairs is thus canceled by a factor of p downstairs, and hence the binomial coefficient has no factor p. $\quad\square$

Theorem A13 (Sylow). *If G is a group of order $p^k m$, where p is a prime not dividing m, then G contains a subgroup of order p^k.*

Proof (Wielandt). If X is the family of all subsets of G of cardinal p^k, then Lemma A12 shows that $p \nmid |X|$. Let G act on X by left translation: if $B \subset G$ and $|B| = p^k$, then

$$g \cdot B = \{gb : b \in B\}.$$

There is some orbit $o(B)$ with $p \nmid |o(B)|$ (otherwise p divides the cardinal of every orbit, hence p divides $|X|$). Choose such a subset $B \in X$. Now $|G|/|G_B| = [G : G_B] = |o(B)|$ is prime to p; it follows that $|G_B| = p^k m' \geq p^k$ for some $m' \mid m$. On the other hand, if $b_0 \in B$ and $g \in G_B$, then $gb_0 \in g \cdot B = B$ (definition of stabilizer); moreover, if g, h are distinct elements of G_B, then gb_0 and hb_0 are distinct elements of B. Therefore $|G_B| \leq |B| = p^k$, and so G_B is a subgroup of order p^k. □

Definition. If $|G| = p^k m$, where p is a prime not dividing m, then a subgroup of G of order p^k is called a **Sylow p-subgroup** of G.

One knows that any two Sylow p-subgroups of a group G are isomorphic (indeed, they are conjugate), and that there are exactly $1 + rp$ of them for some integer $r \geq 0$.

Corollary A14 (Cauchy). *If p is a prime dividing $|G|$, then G contains an element of order p.*

Proof. Let H be a Sylow p-subgroup of G and choose $x \in H^\# = H - \{1\}$. By Lagrange's theorem, the order of x is p^t for some t. If $t = 1$, we are done; if $t > 1$, then it is easy to see that $x^{p^{t-1}}$ has order p. □

Lemma A15. *Every finite abelian group $G \neq \{1\}$ contains a subgroup of prime index.*

Proof. The proof is by induction on the number k of (not necessarily distinct) prime factors of $|G|$. If $k = 1$, then G has prime order and $\{1\}$ has prime index. Assume $k > 1$. By Cauchy's theorem, G contains an element x of prime order; since G is abelian, the cyclic subgroup H generated by x is normal, hence the quotient group G/H is defined. By induction, G/H has a subgroup S^* of prime index, and the correspondence theorem gives a subgroup S of G of prime index. □

Theorem A16. *A group $G \neq \{1\}$ is solvable (it has a normal series with abelian factor groups) if and only if G has a normal series with factor groups of prime order.*

Proof. Sufficiency is obvious; we prove necessity by induction on $|G|$. Assume that

$$G = G_0 \supset G_1 \supset \cdots \supset G_n = \{1\}$$

is a normal series with G_i/G_{i+1} abelian for all i; we may further assume that $G \neq G_1$. By Lemma A15, the abelian group G/G_1 has a (necessarily normal) subgroup S^* of prime index; the correspondence theorem gives an intermediate subgroup S $(G \supset S \supset G_1)$ with S normal in G and with $[G:S] = [G/G_1:S^*]$ prime. Now S is a solvable group (consider the normal series

$$S \supset G_1 \supset G_2 \supset \cdots \supset G_n = \{1\};$$

S/G_1 is abelian because it is a subgroup of the abelian group G/G_1), and induction provides a normal series of it with factor groups of prime order. $\square$

Corollary A17. *Every solvable group has a subgroup of prime index.*

Recall that the **commutator** of elements $x, y \in G$ is

$$[x, y] = xyx^{-1}y^{-1}.$$

The **commutator subgroup** G' of G is the subgroup generated by all the commutators (the product of two commutators may not be a commutator). Note that G' is a normal subgroup of G, for if $a \in G$, then

$$a[x,y]a^{-1} = [axa^{-1}, aya^{-1}];$$

moreover, G/G' is abelian.

Lemma A18. *If H is a normal subgroup of G, then G/H is abelian if and only if $G' \subset H$.*

Proof. If G/H is abelian, then for all $x, y \in G$,

$$xyH = xHyH = yHxH = yxH,$$

and so $xyx^{-1}y^{-1} \in H$; it follows that $G' \subset H$. Conversely, if $G' \subset H$, then the third isomorphism theorem shows that G/H is a quotient group of the abelian group G/G', hence is abelian. $\square$

Definition. The **higher commutator subgroups** are defined inductively:

$$G^{(0)} = G; \qquad G^{(i+1)} = G^{(i)\prime};$$

that is, $G^{(i+1)}$ is the commutator subgroup of $G^{(i)}$.

Lemma A19. *A group G is solvable if and only if $G^{(n)} = \{1\}$ for some n.*

Proof. If G is solvable, then there is a normal series

$$G = G_0 \supset G_1 \supset \cdots \supset G_n = \{1\}$$

with each factor group G_i/G_{i+1} abelian. We prove, by induction on i, that $G_i \supset G^{(i)}$; this will give the result. If $i = 0$, then $G_i = G_0 = G$. Assume, by induction, that $G_i \supset G^{(i)}$; then $G_i' \supset G^{(i)\prime} = G^{(i+1)}$. But G_i/G_{i+1} abelian implies $G_{i+1} \supset G_i'$, by Lemma A18, and so $G_{i+1} \supset G^{(i+1)}$.

Conversely, if $G^{(n)} = \{1\}$, then

$$G = G^{(0)} \supset G' \supset G^{(2)} \supset \cdots \supset G^{(n)} = \{1\}$$

is a normal series with abelian factor groups; hence G is solvable. $\square$

Theorem A20. *If G is a solvable group, then every subgroup and every quotient group of G is also solvable.*

Proof. If H is a subgroup of G, then it is easy to prove by induction that $H^{(i)} \subset G^{(i)}$ for all i. Hence, $G^{(n)} = \{1\}$ implies $H^{(n)} = \{1\}$ and H is solvable.

If $\varphi: G \to K$ is a surjective homomorphism, then $\varphi(G') = K'$: if $uvu^{-1}v^{-1}$ is a commutator in K, choose $x, y \in G$ with $\varphi(x) = u$ and $\varphi(y) = v$; then $\varphi(xyx^{-1}y^{-1}) = uvu^{-1}v^{-1}$. One proves easily, by induction, that $\varphi(G^{(i)}) = K^{(i)}$ for all i. Hence, if G is solvable, then $G^{(n)} = \{1\}$ for some n and $K^{(n)} = \{1\}$; therefore K is solvable. Now take $K = G/N$, where N is any normal subgroup of G, and take φ to be the natural map $G \to G/N$. $\square$

Theorem A21. *Let G be a group with normal subgroup H. If H and G/H are solvable groups, then G is solvable.*

Proof. Let

$$G/H = G^* = G_0^* \supset G_1^* \supset \cdots \supset G_m^* = \{1\}$$

be a normal series with abelian factor groups. By the correspondence theorem, there is a series

$$G = G_0 \supset G_1 \supset \cdots \supset G_m = H$$

with each G_i normal in G_{i-1} and with abelian factor groups. Since H is solvable, there is a normal series

$$H = H_0 \supset H_1 \supset \cdots \supset H_n = \{1\}$$

with abelian factor groups. Splicing these two series together gives a normal series for G with abelian factor groups. $\square$

Definition. The **center** of a group G is

$$Z(G) = \{g \in G: gx = xg \text{ for all } x \in G\}.$$

It is easy to see that $Z(G)$ is an abelian normal subgroup of G. Also, $g \in Z(G)$ if and only if the conjugacy class of g is $\{g\}$, so that $|Z(G)|$ is the number of conjugacy classes of cardinal 1.

There are groups G with $Z(G) = \{1\}$; for example, $Z(S_3) = \{1\}$.

Lemma A22. *If p is a prime and G is a p-group, then $Z(G) \neq \{1\}$.*

Proof. Partition G into its conjugacy classes: using our remark about conjugacy classes of cardinal 1, there is a disjoint union

$$G = Z(G) \cup C_1 \cup \cdots \cup C_t,$$

where the C_i are the conjugacy classes of cardinal larger than 1. If we choose $x_i \in C_i$, then Corollary A11 gives

$$|G| = |Z(G)| + \sum [G \colon C_G(x_i)].$$

By Lagrange's theorem, $[G \colon C_G(x_i)]$ is divisible by p for all i, and so p divides $|Z(G)|$. $\square$

Theorem A23. *Every p-group G is solvable, and hence it has a subgroup of index p if $G \neq \{1\}$.*

Proof. We prove that G is solvable by induction on $|G|$. If $|G| \neq 1$, then $Z(G) \neq \{1\}$, by Lemma A22. If $Z(G) = G$, then G is abelian, hence solvable. If $Z(G) \neq G$, then $G/Z(G)$ is a p-group of order $< |G|$, hence it is solvable, by induction. Since $Z(G)$ is solvable, being abelian, Theorem A21 shows that G is solvable.

The second statement follows from Corollary A17. $\square$

Let us pass from abstract groups to permutation groups; Cayley's theorem shows that this is no loss in generality.

Recall that S_X, the symmetric group on a set X, is the set of all permutations (bijections) of X under composition. If $X = \{x_1, \ldots, x_n\}$, then there is an isomorphism $S_X \xrightarrow{\sim} S_n$ (namely, $\alpha \mapsto \theta \alpha \theta^{-1}$, where $\theta(x_i) = i$) and one usually identifies these two groups.

Theorem A24 (Cayley). *Every group G of order n is (isomorphic to) a subgroup of S_n.*

Proof. If $a \in G$, then the function $\lambda_a \colon G \to G$, defined by $x \mapsto ax$, is a bijection, for its inverse is $\lambda_{a^{-1}} \colon x \mapsto a^{-1}x$; hence $\lambda_a \in S_G \cong S_n$. Define $\lambda \colon G \to S_G$ by $a \mapsto \lambda_a$. It remains to prove that λ is an injective homomorphism.

If $a, b \in G$ are distinct, then $\lambda_a \neq \lambda_b$ (because these two functions have different values on $1 \in G$). Finally, λ is a homomorphism:

$$\lambda_a \lambda_b \colon x \mapsto bx \mapsto a(bx)$$

and

$$\lambda_{ab} \colon x \mapsto (ab)x,$$

so the associative law implies $\lambda_{ab} = \lambda_a \lambda_b$, as desired.   □

Lemma A25. *The alternating group A_n is generated by the 3-cycles.*

Proof. If $\alpha \in A_n$, then $\alpha = \tau_1 \cdots \tau_m$, where each τ_i is a transposition and m is even; hence

$$\alpha = (\tau_1 \tau_2)(\tau_3 \tau_4) \cdots (\tau_{m-1} \tau_m).$$

If τ_{2k-1} and τ_{2k} are not disjoint, then their product is a 3-cycle: $\tau_{2k-1} \tau_{2k} = (ab)(ac) = (acb)$;[1] if they are disjoint, then

$$\tau_{2k-1} \tau_{2k} = (ab)(cd) = (ab)(bc)(bc)(cd) = (bca)(cdb).$$

Therefore α is a product of 3-cycles.   □

Lemma A26. *The commutator subgroup of S_n is A_n.*

Proof. Since S_n/A_n is abelian (it has order 2), Lemma A18 gives $S_n' \subset A_n$. Since A_n is generated by the 3-cycles, it suffices to prove every $\sigma = (ijk)$ is a commutator. Now

$$\sigma^{-1} = \sigma^2 = (ikj) = (ij)(ik),$$

so that

$$\sigma = \sigma^4 = (ij)(ik)(ij)(ik);$$

this is a commutator because $(ij) = (ij)^{-1}$.   □

Lemma A27. *If $\gamma = (i_0, i_1, \ldots, i_{k-1})$ is a k-cycle in S_n and $\alpha \in S_n$, then $\alpha \gamma \alpha^{-1}$ is also a k-cycle; indeed,*

$$\alpha \gamma \alpha^{-1} = (\alpha i_0, \alpha i_1, \ldots, \alpha i_{k-1}).$$

Conversely, if $\gamma' = (i_0', i_1', \ldots, i_{k-1}')$ is another k-cycle, then there exists $\alpha \in S_n$ with $\gamma' = \alpha \gamma \alpha^{-1}$.

Proof. If $\ell \neq \alpha i_j$, $0 \leq j \leq k - 1$, then $\alpha^{-1}\ell \neq i_j$ and so $\gamma(\alpha^{-1}\ell) = \alpha^{-1}\ell$; therefore $\alpha \gamma \alpha^{-1}: \ell \mapsto \alpha^{-1}\ell \mapsto \alpha^{-1}\ell \mapsto \ell$; that is, $\alpha \gamma \alpha^{-1}$ fixes ℓ. If $\ell = \alpha i_j$, then $\alpha \gamma \alpha^{-1}: \ell = \alpha i_j \mapsto i_j \mapsto i_{j+1} \mapsto \alpha i_{j+1}$ (read subscripts mod k). Hence $\alpha \gamma \alpha^{-1}$ and $(\alpha i_0, \alpha i_1, \ldots, \alpha i_{k-1})$ are equal.

---

[1] We multiply permutations from right to left:

$$(\sigma \tau)a = \sigma(\tau(a));$$

that is, $\sigma \tau: a \mapsto \tau a \mapsto \sigma(\tau a)$. In particular, $(ab)(ac) = (acb)$ because

$$(ab)(ac): a \mapsto c \mapsto c; \quad b \mapsto b \mapsto a; \quad c \mapsto a \mapsto b.$$

Conversely, given γ and γ', choose a permutation α with $\alpha i_j = i'_j$ for all j. Then the first part of the proof shows that $\gamma' = \alpha\gamma\alpha^{-1}$.   □

Remark. The same technique proves the lemma with γ a cycle replaced by γ a product of disjoint cycles.

Theorem A28. *The alternating group A_n is the only subgroup of S_n having index 2.*

Proof. We show first that, for any group G, a subgroup H of index 2 must be normal. If $a \in G$, $a \notin H$, then $aH \cap H = \emptyset$ and, by hypothesis, $aH \cup H = G$; hence aH is the complement of H. Since $Ha \cap H = \emptyset$, it follows that $Ha \subset aH$. Now $a \notin H$ and $h \in H$ imply $ha = ah'$ for some $h' \in H$, and so $a^{-1}ha = h' \in H$; therefore H is a normal subgroup of G.

If $[S_n : H] = 2$, then H is normal in S_n, and Lemma A18 gives $A_n = S'_n \subset H$ (for S_n/H has order 2, hence is abelian). But $|A_n| = n!/2 = |H|$, and so $H = A_n$.   □

We are going to prove that A_5 is a simple group.

Lemma A29. (i) *There are 20 3-cycles in S_5, and they are all conjugate in S_5.*
(ii) *All 3-cycles are conjugate in A_5.*

Proof. (i) The number of 3-cycles (abc) is $5 \times 4 \times 3/3 = 20$ (one divides by 3 because $(abc) = (bca) = (cab)$). The conjugacy of any two 3-cycles follows at once from Lemma A27.

(ii) Given 3-cycles γ, γ', one must find an *even* permutation α with $\gamma' = \alpha\gamma\alpha^{-1}$. This can be done directly, but it involves consideration of various cases; here is another proof.

If $\alpha = (123)$ and $C_S(\alpha)$ is the centralizer of α in S_5, then Corollary A11 gives $20 = [S_5 : C_S(\alpha)]$; hence $|C_S(\alpha)| = 6$. But we can exhibit the six elements that commute with α:

$$1, \quad \alpha, \quad \alpha^2, \quad (45), \quad (45)\alpha, \quad (45)\alpha^2.$$

Only the first three of these are even permutations, and so $|C_A(\alpha)| = 3$, where $C_A(\alpha)$ is the centralizer of α in A_5. By Corollary A11, the number of conjugates of α in A_5 is $[A_5 : C_A(\alpha)] = |A_5|/|C_A(\alpha)| = 60/3 = 20$. Therefore, all 3-cycles are conjugate to $\alpha = (123)$ in A_5.   □

Theorem A30. *A_5 is a simple group.*

Proof. If $H \neq \{1\}$ is a normal subgroup of A_5 and if $\sigma \in H$, then every conjugate of σ in A_5 also lies in H. In particular, if H contains a 3-cycle, then it contains all 3-cycles, by Lemma A29(ii); but then $H = A_5$, by Lemma A25.

Let $\sigma \in H$, $\sigma \neq 1$. After a harmless relabeling, we may assume either $\sigma = (123)$, $\sigma = (12)(34)$, or $\sigma = (12345)$ (these are the only possible cycle structures of (even) permutations in A_5). If $\sigma = (123)$, then $H = A_5$, as we have noted above. If $\sigma = (12)(34)$, define $\tau = (12)(35)$; then

$$\tau\sigma\tau^{-1} = (\tau 1\ \tau 2)(\tau 3\ \tau 4) = (12)(45)$$

and

$$\tau\sigma\tau^{-1}\sigma^{-1} = (354) \in H.$$

Finally, if $\sigma = (12345)$, define $\tau = (132)$; then

$$\sigma\tau^{-1}\sigma^{-1} = (\sigma 1\ \sigma 2\ \sigma 3) = (234)$$

and

$$\tau\sigma\tau^{-1}\sigma^{-1} = (134).$$

In each case, H must contain a 3-cycle. Therefore, A_5 contains no proper normal subgroups $\neq \{1\}$ and hence it is simple.   □

Corollary A31. *The only normal subgroups of S_5 are $\{1\}$, A_5, and S_5.*

Proof. Let $H \neq \{1\}$ be a normal subgroup of S_5. The second isomorphism theorem gives $H \cap A_5$ a normal subgroup of A_5; as A_5 is a simple group, either $H \cap A_5 = A_5$ or $H \cap A_5 = \{1\}$. In the first case, $A_5 \subset H$ and $H = A_5$ or $H = S_5$. If $H \cap A_5 = \{1\}$, we claim that $|H| = 2$. If $h\alpha = k\beta$ for $h, k \in H$ and $\alpha, \beta \in A_5$, then $k^{-1}h = \beta\alpha^{-1} \in H \cap A_5 = \{1\}$ and $h = k$ and $\alpha = \beta$; hence S_5 contains $|A_5||H| = 60|H|$ distinct elements. If $h \in H$, $h \neq 1$, then $h = (ab)$ (the only other elements of order 2 have form $(ab)(cd)$ and are even permutations), and it is easy to find a conjugate distinct from h, contradicting the normality of H.   □

Theorem A32. *S_n is solvable for $n \leq 4$, but it is not solvable for $n \geq 5$.*

Proof. If $m < n$, then S_m is (isomorphic to) a subgroup of S_n. Since every subgroup of a solvable group is itself solvable (Theorem A20), it suffices to show that S_4 is solvable and S_5 is not solvable.

Here is a normal series of S_4 that has abelian factor groups:

$$S_4 \supset A_4 \supset V \supset \{1\},$$

where V is the four group (the factor groups have orders 2, 3, 4, respectively, hence are abelian).

Were S_5 solvable, then its subgroup A_5 would also be solvable. Since A_5 is simple, its only normal series is $A_5 \supset \{1\}$, and the (only) factor group is the nonabelian group $A_5/\{1\} \cong A_5$.   □

We now discuss Exercise 97, the group theoretic basis of the computation of the Galois groups of irreducible quartic polynomials over $\mathbb{Q}$.

First of all, we list the subgroups G of S_4 whose order is a multiple of 4: $|G| = 4, 8, 12, 24$. If $|G| = 4$, then the only abstract groups G are $\mathbb{Z}_4$ and $\mathbb{Z}_2 \times \mathbb{Z}_2$, and both occur as subgroups of S_4 (in particular, $V \cong \mathbb{Z}_2 \times \mathbb{Z}_2$). There is a subgroup of order 8 isomorphic to the dihedral group D_8, namely, the symmetries of a square regarded as permutations of the 4 corners; since a subgroup of order 8 is a Sylow 2-subgroup of S_4, all subgroups of order 8 are isomorphic to D_8. Theorem A28 shows that A_4 is the only subgroup of order 12 and, of course, S_4 itself is the only subgroup of order 24.

If $G \subset S_4$ and V is the four group (which is a normal subgroup of S_4), then the second isomorphism theorem gives $G \cap V$ normal in G and $G/G \cap V \cong GV/V \subset S_4/V$. Define

$$m = |G/G \cap V|;$$

it follows that m is a divisor of $[S_4 : V] = 24/4 = 6$. ($S_4/V \cong S_3$, but we do not need this fact.)

Theorem A33 (= Exercise 97). *Let $G \subset S_4$ have order a multiple of* 4 *and let $m = |G/G \cap V|$. If $m = 6$, then $G = S_4$; if $m = 3$, then $G = A_4$; if $m = 1$, then $G = V$; if $m = 2$, then $G \cong D_8$ or $\mathbb{Z}_4$ or V.*

Proof. If $m = 6$ or 3, then $|G| \geq 12$ ($|G|$ is divisible by 3 and, by hypothesis, 4). By Theorem A28, A_4 is the only subgroup of S_4 of order 12, and so $A_4 \subset G$ in either case. But $V \subset A_4$. It follows easily that $m = 6$ forces $G = S_4$ and $m = 3$ forces $G = A_4$.

If $m = 1$, then $G = G \cap V$ and $G \subset V$; since $|G|$ is a multiple of 4, it follows that $G = V$.

If $m = 2$, then $|G| = 2|G \cap V|$; since $|V| = 4$, we have $|G \cap V| = 1, 2$, or 4. We cannot have $|G \cap V| = 1$ lest $|G| = 2$, which is not a multiple of 4. If $|G \cap V| = 4$, then $|G| = 8$ and $G \cong D_8$ (as we remarked above, D_8 is a Sylow 2-subgroup). If $|G \cap V| = 2$, then $|G| = 4$ and $G \cong \mathbb{Z}_4$ or V (these are the only abstract groups of order 4). $\square$

The possibility $m = 2$ and $G \cong V$ can occur. Let G be the following isomorphic copy of V in S_4:

$$G = \{1, (12)(34), (12), (34)\}.$$

Note that $G \cap V = \{1, (12)(34)\}$ and $m = |G/G \cap V| = 4/2 = 2$. This group G does not act transitively on $\{1, 2, 3, 4\}$ because, for example, there is no $g \in G$ with $g(1) = 3$. Exercise 98 invokes the extra hypothesis of G acting transitively to eliminate the case $G \cong V$ from the list of candidates for G when $m = 2$.

Lemma A34. *If G is a group and H is a subgroup of index n, then there is a homomorphism $\varphi: G \to S_n$ with $\ker \varphi \subset H$.*

Proof. Let X be the family of all cosets of H in G; since $|X| = n$, it is easy to see that $S_X \cong S_n$ (where S_X is the group of all permutations of X). For $g \in G$, define $\varphi(g): X \to X$ by $\varphi(g): aH \mapsto gaH$ (where $a \in G$); note that $\varphi(g)$ is a bijection, for its inverse is $\varphi(g^{-1})$. To see that φ is a homomorphism, compute

$$\varphi(gg'): aH \mapsto (gg')aH;$$
$$\varphi(g)\varphi(g'): aH \mapsto g'aH \mapsto g(g'aH).$$

If $\varphi(g)$ is the identity on X, then $\varphi(g): aH \mapsto aH$ for all $a \in G$; in particular, $\varphi(g): H \mapsto H$, so that $gH = H$ and $g \in H$. $\square$

Theorem A35. A_6 *has no subgroups of prime index.*

Proof. It is known that A_6 is a simple group of order $360 = 2^3 \cdot 3^2 \cdot 5$. If H is a subgroup of prime index, then $[A_6 : H] = 2$, 3, or 5. By Lemma A34, there is a homomorphism $\varphi: A_6 \to S_n$, where $n = 2$, 3, or 5, with $\ker \varphi \subset H$; in particular, $\ker \varphi$ is a normal subgroup of A_6 with $\ker \varphi \neq A_6$. Since A_6 is simple, $\ker \varphi = \{1\}$ and φ is an injection. But this is impossible because $|S_5| = 120 < 360$. $\square$

Lemma A36. S_5 *has no subgroups of order* 30 *or of order* 40.

Proof. Suppose H is a subgroup of order 30; that is, H has index $[S_5 : H] = 120/30 = 4$. Lemma A34 gives a homomorphism $\varphi: S_5 \to S_4$ with $\ker \varphi \subset H$. But $\ker \varphi$ is a normal subgroup of S_5, and so its order must be 1, 60, or 120 (Corollary A31). Since $|H| = 30$, it follows that $\ker \varphi = \{1\}$, and S_5 is isomorphic to a subgroup of S_4, a contradiction. A similar argument shows that S_5 has no subgroup of index 3. $\square$

Theorem A37. *If α is a 5-cycle in S_5 and τ is a transposition in S_5, then* $\langle \alpha, \tau \rangle = S_5$.

Proof. Let $H = \langle \alpha, \tau \rangle$ be the subgroup generated by α and τ. We may assume that $\alpha = (1\,2\,3\,4\,5)$ and $\tau = (1\,i)$. Now some power of α, say, α^k carries i into 1, so that Lemma A27 gives $\alpha^k(1\,i)\alpha^{-k} = (j\,1)$ for some j (actually, $j = \alpha^k 1$). Note that $i \neq j$ because $(1i)$ is not a disjoint factor of the 5-cycle α^k. But $(1i)(1j) = (1\,j\,i)$, an element of order 3. The order of H is thus divisible by 2, 3, and 5, hence $|H| \geq 30$. By Lemma A36, $|H| = 60$ or 120. If $|H| = 60$, then $H = A_5$, by Theorem A28; but $H \neq A_5$ because $\tau \in H$ is an odd permutation. Therefore $H = S_5$. $\square$

(A more computational proof shows first that every transposition can be obtained from α and τ, and then that S_5 is generated by the transpositions.)

Theorem A38. *A subgroup H of S_5 is solvable if and only if $|H| \leq 24$.*

Proof. We leave to the reader the fact that every group of order ≤ 24 is solvable (whether or not it is a subgroup of S_5; indeed, every group of order < 60 is solvable).

Since $|S_5| = 120$, the only divisors of $|S_5|$ larger than 24 are 30, 40, 60, and 120. Now S_5 itself is not solvable, by Theorem A32; also, A_5 is the only subgroup of order 60 (Theorem A28), and it is not solvable because it is simple and not abelian (Theorem A30). Lemma A36 completes the proof. □

Theorem A38 is used in Exercise 102. It is implicit in the second part of this exercise that S_5 does have a subgroup of order 20; the normalizer of a Sylow 5-subgroup is such a subgroup. Of course, S_5 does have a solvable subgroup of order 24, namely, S_4.

Appendix 3
Ruler-Compass
Constructions

We are going to show that the classical Greek problems: squaring the circle, duplicating the cube, and trisecting an angle, are impossible to solve. As we shall see, the discussion uses only elementary field theory; no Galois theory is required.

It is essential to state the problems carefully and to agree on certain ground rules. For example, it is clear one can trisect a 60° angle with a protractor (or any other device than can measure an angle); after all, one can divide any number by 3. However, the Greek problems specify that only two tools are allowed, and each must be used in only one way. Let P and Q be points in the plane; we denote the line segment with endpoints P and Q by PQ, and we denote the length of this segment by $|PQ|$. A *ruler* (or *straight-edge*) is a tool that can draw the line $L(P,Q)$ determined by P and Q; a *compass* is a tool that draws the circle with radius $|PQ|$ and center either P or Q; denote these circles by $C(P,Q)$ or $C(Q,P)$, respectively. Since every construction has only a finite number of steps, we shall be able to define "constructible" points inductively.

Given the plane, we establish a coordinate system by first choosing two distinct points, A and B; call the line they determine the *x-axis*. Use a compass to draw the two circles of radius $|AB|$ with centers A and B, respectively. These two circles intersect in two points; the line they determine is called the *y-axis;* it is the perpendicular bisector of AB, and it intersects the x-axis in a point O, called the *origin*. We define the distance $|OA|$ to be 1. We have introduced coordinates in the plane; in particular, $A = (1,0)$ and $B = (-1,0)$.

Definition. If X is a set of points, then a point Q is *0-constructible over* X if $Q \in X$; a point Q is *n-constructible over* X if there are points $P_1, \ldots, P_{n-1}$ such that:

(1) each P_i is $k(i)$-constructible over X with $k(i) < n$;

(2) there are (not necessarily distinct) points E, F, G, H in $X \cup \{P_1, \ldots, P_{n-1}\}$ with either

   (i) Q being the intersection of $L(E, F)$ and $L(G, H)$,

  (ii) Q being an intersection of $L(E, F)$ and $C(G, H)$, or

 (iii) Q being an intersection of $C(E, F)$ and $C(G, H)$.

A point Q is *constructible over* X if it is n-constructible over X for some $n \geq 0$. Finally, a point Q is (absolutely) **constructible** if it is constructible over $X = \{A = (1, 0), B = (-1, 0)\}$.

Euclid proved that every line segment PQ can be bisected; according to our definition, the midpoint M is constructible over $X = \{P, Q\}$. It is plain, however, that there are only countably many (absolutely) constructible points (once the distinct points A and B are chosen).

In our discussion, we shall freely use any standard result of euclidean geometry. Furthermore, we shall identify the complex number $z = x + iy$ with the point (x, y); in particular, we may speak of constructible complex numbers and constructible real numbers.

Lemma B1. *A point* $P = (x, y)$ *is constructible if and only if* $(x, 0)$ *and* $(y, 0)$ *are constructible points. Therefore, a complex number* $z = x + iy$ *is constructible if and only if its real part* x *and its imaginary part* y *are constructible.*

Proof. The point $(x, 0)$ is constructible, for it is the intersection of the x-axis and the vertical line through P; similarly, $(0, y)$ is constructible. But $(y, 0)$ is the intersection of the x-axis and the circle of radius y and center the origin 0, hence is constructible.

Conversely, assume that $(x, 0)$ and $(y, 0)$ are constructible. The point $(0, y)$ is constructible, being the intersection of the y-axis and the circle of radius y and center the origin. One can draw the vertical line at $(x, 0)$ as well as the horizontal line at $(0, y)$, and (x, y) is the intersection of these lines. $\square$

Lemma B2. *Let* K *denote the set of all constructible numbers. Then* K *is a subfield of* $\mathbb{C}$ *if and only if* $K \cap \mathbb{R}$ *is a subfield of* $\mathbb{R}$. *Moreover, if* $K \cap \mathbb{R}$ *is a subfield in which positive elements have square roots, then* K *is closed under square roots.*

Proof. Necessity is clear. Conversely, assume $K \cap \mathbb{R}$ is a subfield of $\mathbb{R}$ and $z = a + ib$, $w = c + id \in K$. By Lemma B1, $a, b, c, d \in K \cap \mathbb{R}$. Now $z + w = a + c + i(b + d)$. By hypothesis, $a + c, b + d \in K \cap \mathbb{R}$, and so Lemma B1 gives $z + w \in K$. Similar arguments show that $zw \in K$ and $z^{-1} \in K$.

Suppose that $K \cap \mathbb{R}$ is a subfield whose elements have square roots in $K \cap \mathbb{R}$. If $z = x + iy \in K$, then $r^2 = x^2 + y^2 \in K \cap \mathbb{R}$, and hence $r \in K \cap \mathbb{R}$.

Now write z in polar form: $z = re^{i\theta}$. The hypothesis gives $\sqrt{r} \in K \cap \mathbb{R}$, and $e^{i\theta/2}$ is constructible because every angle can be bisected. □

Theorem B3. *The set of all constructible numbers K is a subfield of $\mathbb{C}$ that is closed under square roots.*

Proof. It suffices to prove the properties of $K \cap \mathbb{R}$ in Lemma B2. Let a, b be constructible reals.

(i) $-a$ is constructible.

If $P = (a, 0)$ is a constructible point, then $(-a, 0)$ is the other intersection of the x-axis and $C(0, P)$.

(ii) $a + b$ is constructible.

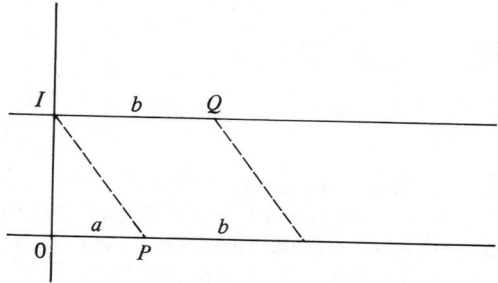

Let $I = (0, 1)$, $P = (a, 0)$ and $Q = (1, b)$; it is easy to see that Q is constructible. The line through Q parallel to IP intersects the x-axis in $(a + b, 0)$, as desired. Although the picture is drawn with a, b positive, it is clear that this construction works for any choice of signs for a, b.

(iii) ab is constructible. We may assume $a > 0$ and $b > 0$.

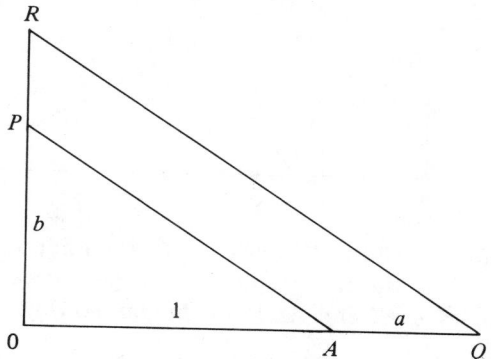

In this picture, $A = (1, 0)$, $Q = (1 + a, 0)$, and $P = (0, b)$. Define R to be the intersection of the y-axis and the line through Q parallel to AP. Since the triangles OAP and OQR are similar,

$$|OQ|/|OA| = |OR|/|OP|;$$

hence $(a+1)/1 = (b+|PR|)/b$, and $|PR| = ab$. It follows that $(ab,0)$ is constructible.

(iv) a^{-1} is constructible.

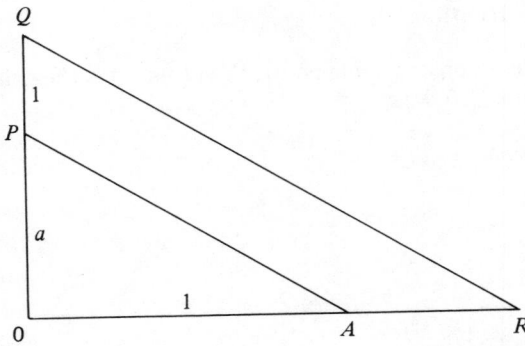

The construction is as that in (iii), but with a different selection of points. Let $A = (1,0)$, $P = (0,a)$, and $Q = (0,1+a)$. Define R as the intersection of the x-axis and the line through Q parallel to AP. As in (iii), similar triangles may be used to obtain $|AR| = a^{-1}$. Therefore, $(a^{-1},0)$ is constructible.

(v) $\sqrt{a}$ is constructible.

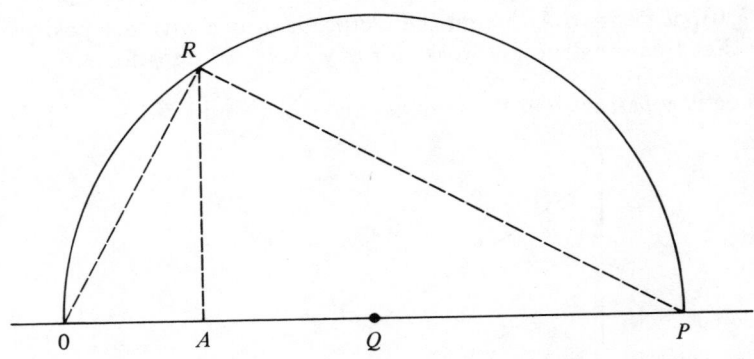

Let $A = (1,0)$ and $P = (1+a,0)$; construct Q, the midpoint of OP. Define R as the intersection of the circle $C(Q,0)$ with the vertical line through A. The (right) triangles AOR and ARP are similar, so that

$$|OA|/|AR| = |AR|/|AP|,$$

and so $|AR| = \sqrt{a}$.   □

Since every subfield of $\mathbb{C}$ contains $\mathbb{Q}$, it follows that K contains all points (x, y) with both coordinates rational.

Corollary B4. *If a, b, c are constructible, then the roots of the quadratic $ax^2 + bx + c$ are also constructible.*

Proof. This follows from the theorem and the quadratic formula. $\square$

Recall that if E is a subfield of $\mathbb{C}$ (necessarily) containing $\mathbb{Q}$, then E may be regarded as a vector space over $\mathbb{Q}$; its dimension is denoted by $[E:\mathbb{Q}]$. If $z \in \mathbb{C}$, then $\mathbb{Q}(z)$ is the smallest subfield of $\mathbb{C}$ that contains z.

Lemma B5. *Let k be a subfield of $\mathbb{C}$. (i) A line with equation $y = ax + b$ or $x = b$ is of the form $L(E, F)$ for $E, F \in k$ if and only if $a, b \in k$.*
(ii) A circle with equation $(x - a)^2 + (y - b)^2 = c^2$ is of the form $C(E, F)$ for $E, F \in k$ if and only if $a, b, c \in k$.

Proof. If $E = (p, q)$ and $F = (s, t)$ lie in k, the equations of $L(E, F)$ and $C(E, F)$ express a, b, c in terms of rational combinations and square roots of p, q, s, t in k. Conversely, the line contains the points $E = (0, b)$ and $F = (-b/a, 0)$ in k (in the second case, $E = (b, 0)$ and $F = (b, 1)$). The circle of the given equation is $C(E, F)$, where $E = (a, b)$ and $F = (a, b + c)$. $\square$

Theorem B6. *A complex number z is constructible if and only if there is a tower of fields*

$$\mathbb{Q} = K_0 \subset K_1 \subset \cdots \subset K_n,$$

where $[K_{i+1}:K_i] \leq 2$ for all i, and $z \in K_n$.

Proof. If z is constructible, there is a sequence of points $1, -1, z_1, \ldots, z_n = z$ with each z_i obtainable from $\{1, -1, z_1, \ldots, z_{i-1}\}$. Define $K_i = \mathbb{Q}(z_1, \ldots, z_i)$. Given $u = z_{i+1}$, there are points $E, F, G, H \in K_i$ with one of the following:

$$u = L(E, F) \cap L(G, H); \quad u \in L(E, F) \cap C(G, H); \quad u \in C(E, F) \cap C(G, H).$$

The reader may compute the coordinates of the point(s) of intersection to see that, in the first case, $u \in K_i$, while in the other two cases, u is a root of a quadratic polynomial over K_i.

Conversely, given the tower of fields, choose $u_i \in K_i$ with $K_{i+1} = K_i(u_{i+1})$ for all i. We prove, by induction, that every element of K_i is constructible. The induction begins, for $K_0 = \mathbb{Q}$. Now every element $v \in K_{i+1}$ is a root of some quadratic polynomial over K_i, and so Corollary B4 gives v constructible. $\square$

Corollary B7. *If a complex number z is constructible, then $[\mathbb{Q}(z):\mathbb{Q}]$ is a power of 2.*

Proof. This follows from the theorem and Lemma 31. □

Remark. The converse of this corollary is false. For every $m \geq 2$, there exists an irreducible polynomial $p(x) \in \mathbb{Q}[x]$ of degree $n = 2^m$ which has Galois group $G = \mathrm{Gal}(E/\mathbb{Q}) \cong S_n$ (see [Hadlock, p. 218]). For example, we showed in the text that $x^4 - 4x + 2$ is irreducible over $\mathbb{Q}$ and has Galois group S_4. Were every root of $p(x)$ constructible, then Theorem B3 would imply that every element of E is constructible. If H is a Sylow 2-subgroup of G, however, then $[G:H]$ is an odd number; the intermediate field E^H thus has odd degree ($[E^H:\mathbb{Q}] = [G:H]$) and so none of its elements are constructible, by Corollary B7. This contradiction shows that some root of $p(x)$ is not constructible (yet every root has degree 2^m over $\mathbb{Q}$).

It is now a simple matter to dispose of some famous problems.

(1) It is impossible to "square the circle."

The problem is to construct, with ruler and compass, a square whose area is equal to the area of a circle of radius 1; in other words, one asks whether $\sqrt{\pi}$ is constructible. But it is a classical result that π, hence $\sqrt{\pi}$, is transcendental (over $\mathbb{Q}$) (see [Hadlock, p. 47]), and so it does not lie in any finite extension of $\mathbb{Q}$, let alone one of degree a power of 2.

(2) It is impossible to "duplicate the cube."

The problem is to construct a cube whose volume is 2; in other words, is the cube root of 2, call it α, constructible? Now $x^3 - 2$ is irreducible over $\mathbb{Q}$, by Eisenstein, and so $[\mathbb{Q}(\alpha):\mathbb{Q}] = 3$, which is not a power of 2. Corollary B7 gives the result.

(3) It is impossible to trisect an arbitrary angle.

An angle θ is given by two intersecting lines; it is no loss in generality to assume the lines intersect at the origin and that one line is the x-axis. If we could draw the angle trisector, then we could construct the point $(\cos\theta/3, \sin\theta/3)$ which is the intersection of the trisector and the unit circle; hence $\cos\theta/3$ would be constructible, by Lemma B1. Now some angles, e.g., $\theta = \pi/2$, can be trisected. On the other hand, we will now show that $\theta = \pi/3$ cannot be trisected. Computing the real part of $e^{3i\theta} = (\cos\theta + i\sin\theta)^3$ gives the trigonometric identity:

$$\cos 3\theta = 4\cos^3\theta - 3\cos\theta.$$

Defining $u = 2\cos\theta$ and $\theta = \pi/9$, we arrive at the equation

$$u^3 - 3u - 1 = 0.$$

It is easy to see that this cubic is irreducible (it has no rational root; now use Exercise 70), and so $[\mathbb{Q}(u):\mathbb{Q}] = 3$. Corollary B7 shows that u is not constructible.

(4) Regular p-gons

Galois theory will be used in discussing this problem.

Theorem B8 (Gauss). *If p is an odd prime, then a regular p-gon is constructible if and only if $p = 2^{2^t} + 1$ for some $t \geq 0$.*

Proof. This is again a question of constructibility of a point on the unit circle, namely, $z = e^{2\pi i/p}$. Now the irreducible polynomial of z over $\mathbb{Q}$ is the cyclotomic polynomial $\Phi_p(x)$ of degree $p - 1$ (Corollary 25).

Assume z is constructible. By Corollary B7, $p - 1 = 2^s$ for some s. We claim that s itself is a power of 2. Otherwise, there is an odd number k with $s = km$. But $x^k + 1$ factors over $\mathbb{Z}$ (because -1 is a root); setting $x = 2^m$ thus gives a forbidden factorization of p.

Conversely, assume $p = 2^{2^t} + 1$ is prime. Since z is a primitive pth root of unity, $\mathbb{Q}(z)$ is the splitting field of $\Phi_p(x)$ over $\mathbb{Q}$. Therefore $|\text{Gal}(\mathbb{Q}(z)/\mathbb{Q})| = 2^{2^t}$ and the Galois group is a 2-group. But a 2-group has a normal series in which each factor has order 2 (this follows easily from Theorem A23); by the fundamental theorem of Galois theory, there is a tower of fields $\mathbb{Q} \subset K_1 \subset \cdots \subset K_m = \mathbb{Q}(z)$ with $[K_{i+1} : K_i] = 2$ for all i, that is, z is constructible. $\square$

Gauss actually gave a geometric construction of the regular 17-gon.

Remark. Primes of the form $2^{2^t} + 1$ are called **Fermat primes**. The values $0 \leq t \leq 4$ do give primes (they are 3, 5, 17, 257, 65,537), the next few values of t do not give primes, and it is unknown whether any other Fermat primes exist.

Corollary B9. *It is impossible to construct a regular 7-gon, a regular 11-gon, or a regular 13-gon.*

Proof. 7, 11, and 13 are not Fermat primes. $\square$

The following result is known (see [Hadlock, p. 106]):

Theorem. *A regular n-gon is constructible if and only if n is a product of a power of 2 and distinct Fermat primes.*

It follows that regular 9-gons and regular 14-gons are not constructible; on the other hand, a regular 15-gon is constructible.

Appendix 4
Old-fashioned Galois
Theory

Gimme that old-time Galois theory;

If it's good enough for Galois, then it's good enough for me!

I am a creature of the twentieth century; algebraic systems and their automorphism groups are part of my mother's milk. When writing the definition of Galois group for this text, I asked an obvious question: how did such
thoughts occur to Galois in the late 1820's? The answer, of course, is that
he did not think in such terms; for its first century, 1830–1930, the Galois
group was a group of permutations. In the late 1920's, E. Artin, developing
ideas of E. Noether going back at least to Dedekind, recognized that it is
both more elegant and more fruitful to describe Galois groups in terms of
field automorphisms (Artin's version is isomorphic to the original version).
In 1930, van der Waerden incorporated much of Artin's viewpoint into his
influential text "Moderne Algebra," and a decade later Artin published his
own lectures. So successful have Artin's ideas proved to be that they have
virtually eclipsed earlier expositions. But we have lost the inevitability of
the definition; group theory is imposed on the study of polynomials rather
than arising naturally from it. This appendix is an attempt to remedy this
pedagogical problem by telling the story of what happened in the beginning. The reader interested in a more thorough account may read [Edwards]
or [Tignol].

We use modern notation and terms even though they were unknown
in the eighteenth century. In particular F shall denote a subfield of the
complex numbers. Permutations arise simultaneously with the question of
finding the roots of a polynomial. If

$$f(x) = \prod_{i=1}^{n}(x - \alpha_i) = x^n + b_{n-1}x^{n-1} + \cdots + b_1 x + b_0,$$

then one sees easily that b_{n-j} is, to sign, the sum of all products of j roots α_i:

$$b_{n-j} = (-1)^j \sum_{1 \le i_1 < i_2 < \cdots < i_j \le n} \alpha_{i_1} \alpha_{i_2} \cdots \alpha_{i_j}.$$

Thus

$$b_{n-1} = -\sum \alpha_i = -(\alpha_1 + \cdots \alpha_n)$$

$$b_{n-2} = \sum_{i<j} \alpha_i \alpha_j$$

$$b_{n-3} = -\sum_{i<j<k} \alpha_i \alpha_j \alpha_k$$

$$\vdots$$

$$b_0 = (-1)^n \alpha_1 \alpha_2 \cdots \alpha_n.$$

Since the coefficients b_{n-j} are unchanged if the roots are re-indexed, it is clear that they are symmetric functions of the roots in the following sense.

Definition. A polynomial $g(x_1, \ldots, x_n) \in F[x_1, \ldots, x_n]$ is **symmetric** if

$$g(x_{\sigma 1}, \ldots, x_{\sigma n}) = g(x_1, \ldots, x_n)$$

for every $\sigma \in S_n$.

Each of the polynomials

$$e_j(x_1, \ldots, x_n) = \sum_{1 \le i_1 < i_2 < \cdots < i_j \le n} x_{i_1} x_{i_2} \cdots x_{i_j}$$

is symmetric; one calls $e_1, \ldots, e_n$ the **elementary symmetric functions.** Note that $e_j(\alpha_1, \ldots, \alpha_n) = (-1)^j b_{n-j}$.

The following result was well known.

Fundamental Theorem of Symmetric Functions. *If $g(x_1, \ldots, x_n) \in F[x_1, \ldots, x_n]$ is symmetric, then there exists $h(x_1, \ldots, x_n) \in F[x_1, \ldots, x_n]$, not necessarily symmetric, with*

$$g(x_1, \ldots, x_n) = h(e_1, \ldots, e_n).$$

For a proof, see [Hadlock, p. 42]. In 1770, Waring published an algorithm for finding h. For example,

$$x_1^2 + x_2^2 + x_3^2 = (x_1 + x_2 + x_3)^2 - 2(x_1 x_2 + x_1 x_3 + x_2 x_3)$$
$$= e_1^2 - 2e_2.$$

Corollary. *Let* $f(x) = x^n + b_{n-1}x^{n-1} + \cdots + b_1 x + b_0 \in F[x]$ *have* (*complex*) *roots* $\alpha_1, \ldots, \alpha_n$; *if* $g(x_1, \ldots, x_n) \in F[x_1, \ldots, x_n]$ *is symmetric, then*

$$g(\alpha_1, \ldots, \alpha_n) \in F.$$

Proof. By the fundamental theorem, there is $h(x_1, \ldots, x_n) \in F[x_1, \ldots, x_n]$ with $g(x_1, \ldots, x_n) = h(e_1, \ldots, e_n)$. Evaluating at $(x_1, \ldots, x_n) = (\alpha_1, \ldots, \alpha_n)$ gives $g(\alpha_1, \ldots, \alpha_n) = h(-b_{n-1}, \ldots, \pm b_0) \in F$. $\square$

The classical formulas for the roots of cubics and quartics, discovered more than two centuries earlier, were also well known. Recall that the roots of $f(x) = x^3 + qx + r$ are:

$$\alpha_1 = y + z; \quad \alpha_2 = \omega y + \omega^2 z; \quad \alpha_3 = \omega^2 y + \omega z$$

(here, ω is a primitive cube root of unity, $y^3 = \frac{1}{2}(-r + \sqrt{R})$, $z^3 = \frac{1}{2}(-r - \sqrt{R})$, and $R = r^2 + 4q^3/27$). In 1770, Lagrange and Vandermonde, independently, sought to find the basic principles underlying the known formulas. They expressed the radicals in terms of the roots α_i:

$$3y = \alpha_1 + \omega\alpha_2 + \omega^2\alpha_3;$$
$$3z = \alpha_1 + \omega^2\alpha_2 + \omega\alpha_3.$$

For given $\alpha_1, \alpha_2, \alpha_3$ and (not necessarily primitive) cube root of unity ζ, let us denote

$$\psi(\zeta) = (\alpha_1 + \zeta\alpha_2 + \zeta^2\alpha_3)^3;$$

then

$$(3y)^3 = \psi(\omega) \quad \text{and} \quad (3z)^3 = \psi(\omega^2)$$

and, for $i = 1, 2, 3$,

$$\alpha_i = \frac{1}{3}\left(\sqrt[3]{\psi(\omega)} + \sqrt[3]{\psi(\omega^2)} \right)$$

if one selects cube roots properly.

How can one determine the two numbers $\psi(\omega)$ and $\psi(\omega^2)$? Regard the roots $\alpha_1, \alpha_2, \alpha_3$ as indeterminates and define:

$$\varphi_1(x_1, x_2, x_3) = x_1 + \omega x_2 + \omega^2 x_3;$$
$$\varphi_2(x_1, x_2, x_3) = x_1 + \omega^2 x_2 + \omega x_3.$$

Neither φ_1 nor φ_2 is symmetric. But the transposition (23) interchanges φ_1 and φ_2 ((23) sends $x_1 \mapsto x_1$, $x_2 \mapsto x_3$, and $x_3 \mapsto x_2$) and the 3-cycle (132) fixes both φ_1^3 and φ_2^3 (for example, (132) sends φ_1^3 into $(x_3 + \omega x_1 + \omega^2 x_2)^3 = [\omega(\omega^2 x_3 + x_1 + \omega x_2)]^3 = \varphi_1^3$; this is one reason for cubing φ_1 and φ_2). It follows that $\varphi_1^3 + \varphi_2^3$ and $\varphi_1^3\varphi_2^3$ are symmetric functions (each is invariant under (23) and (132), and these two permutations generate the symmetric group S_3). The algorithm for the fundamental theorem of symmetric functions expresses $\varphi_1^3 + \varphi_2^3$ and $\varphi_1^3 \varphi_2^3$ in terms of elementary symmetric functions. Since $\varphi_1(\alpha_1, \alpha_2, \alpha_3)^3 = \psi(\omega)$ and $\varphi_2(\alpha_1, \alpha_2, \alpha_3)^3 = \psi(\omega^2)$, the corollary of the fundamental theorem expresses $b_1 = \psi(\omega) + \psi(\omega^2)$ and $b_0 = \psi(\omega)\psi(\omega^2)$ in terms of the coefficients q and r of $f(x)$. We have seen that once we know $\psi(\omega)$ and $\psi(\omega^2)$, we can find the roots $\alpha_1, \alpha_2, \alpha_3$ of $f(x)$. But

$$x^2 - b_1 x + b_0 = (x - \psi(\omega))(x - \psi(\omega^2)),$$

and so $\psi(\omega)$ and $\psi(\omega^2)$ can be found by the quadratic formula. (There are four more polynomials obtained from $\varphi_1(x_1, x_2, x_3)$ by permuting variables: $\omega \varphi_1$; $\omega^2 \varphi_1$; $\omega \varphi_2$, $\omega^2 \varphi_2$. These are the other cube roots of $\psi(\omega)$ and $\psi(\omega^2)$, and so they enter the cubic formula when the condition $yz = -q/3$ is imposed.)

Both Lagrange and Vandermonde did a similar analysis of the quartic. If the roots are $\alpha_1, \alpha_2, \alpha_3, \alpha_4$, then they defined

$$\varphi_1(x_1, x_2, x_3, x_4) = x_1 + i x_2 + i^2 x_3 + i^3 x_4$$

where $i^2 = -1$, and they showed that φ_1^4 plays a decisive role in obtaining the classical formula.

Lagrange generalized this analysis to polynomials $f(x)$ of degree n. If ζ is an nth root of unity (not necessarily primitive) and if the roots of $f(x)$ are $\alpha_1, \ldots, \alpha_n$, first define

$$\varphi_1(\zeta) = \alpha_1 + \alpha_2 \zeta + \alpha_3 \zeta^2 + \cdots + \alpha_n \zeta^{n-1},$$
$$\varphi_2(\zeta) = \alpha_2 + \alpha_3 \zeta + \alpha_4 \zeta^2 + \cdots + \alpha_1 \zeta^{n-1},$$
$$\vdots$$
$$\varphi_n(\zeta) = \alpha_n + \alpha_1 \zeta + \alpha_2 \zeta^2 + \cdots + \alpha_{n-1} \zeta^{n-1},$$

and then define

$$\psi(\zeta) = \varphi_1(\zeta)^n.$$

Of course, we can evaluate $\psi(1) = (\alpha_1 + \alpha_2 + \cdots + \alpha_n)^n$ because the sum of the α_i is, to sign, the coefficient of x^{n-1} in $f(x)$.

Lemma. *If $f(x)$ has degree n, than the roots $\alpha_1, \ldots, \alpha_n$ of $f(x)$ are determined by the $n - 1$ numbers $\psi(\omega)$, $\psi(\omega^2), \ldots, \psi(\omega^{n-1})$, where ω is a primitive nth root of unity.*

Proof. For fixed k with $1 \leq k \leq n - 1$, we have

$$\sum_{j=0}^{n-1} \omega^{kj} = 0$$

(this geometric series sums to $\frac{1-(\omega^k)^n}{1-\omega^k} = 0$ because $\omega^{kn} = 1$ and $\omega^k \neq 1$). Thus,

$$\sum_{j=0}^{n-1} \varphi_1(\omega^j) = \sum_j (\alpha_1 + \alpha_2\omega^j + \alpha_3\omega^{2j} + \cdots + \alpha_n\omega^{(n-1)j})$$

$$= n\alpha_1 + \alpha_2\sum\omega^j + \alpha_3\sum\omega^{2j} + \cdots + \alpha_n\sum\omega^{(n-1)j}$$

$$= n\alpha_1,$$

by our first remark. Similarly, $\sum_j \varphi_i(\omega^j) = n\alpha_i$. If we know $\psi(\omega^j) = \varphi_1(\omega^j)^n$, then we know $\varphi_1(\omega^j)$ as well as $\zeta\varphi_1(\omega^j)$ for every nth root of unity ζ. But it is easy to see that

$$\varphi_i(\zeta) = \zeta^{-i-1}\varphi_1(\zeta)$$

for every ζ. Therefore the roots $\alpha_1, \ldots, \alpha_n$ are determined by $\psi(\omega)$, $\psi(\omega^2), \ldots, \psi(\omega^{n-1})$. $\square$

This lemma, essentially due to Bézout (1765), says that the n roots of $f(x)$, a polynomial of degree n, can be found in terms of $n-1$ numbers $\psi(\omega), \ldots, \psi(\omega^{n-1})$; that is, there is a polynomial of degree $n-1$, namely,

$$\rho(x) = \prod_j (x - \psi(\omega^j)),$$

whose roots determine the roots of $f(x)$. Does this not give the inductive step for finding the roots of a polynomial of arbitrary degree n? The answer, unfortunately, is negative because we do not know the coefficients of $\rho(x)$.[1] At the very least, we need these coefficients to lie in F; and it is precisely this that introduces groups into the theory! Lagrange's idea was to replace $\rho(x)$ by a more manageable polynomial in $F[x]$.

The number $\psi(\omega) = (\alpha_1 + \alpha_2\omega + \cdots + \alpha_n\omega^{n-1})^n$ is not symmetric in the α; let us try to force it to be. If $g(x_1, \ldots, x_n) \in F[x_1, \ldots, x_n]$ and $\sigma \in S_n$, define a polynomial σg by

$$\sigma g(x_1, \ldots, x_n) = g(x_{\sigma 1}, \ldots, x_{\sigma n});$$

just permute the indeterminates as σ prescribes. Consider the "symmetrized" polynomial

$$g^*(x) = \prod_{\sigma \in S_n} (x - \sigma g(x_1, \ldots, x_n));$$

its coefficients have the form

$$e(\sigma_1 g(x_1, \ldots, x_n), \ldots, \sigma_{n!} g(x_1, \ldots, x_n)),$$

---

[1] We gave an argument above that these coefficients are known when $n = 3$.

where e is an elementary symmetric function (of the $n!$ terms $\sigma g(x_1, \ldots, x_n)$) and the permutations in S_n are listed $\sigma_1, \sigma_2, \ldots, \sigma_{n!}$. If τ is any permutation in S_n, then

$$e(\sigma_1 g(x_{\tau 1}, \ldots, x_{\tau n}), \ldots, \sigma_{n!} g(x_{\tau 1}, \ldots, x_{\tau n}))$$
$$= e(\sigma_1 \tau g(x_1, \ldots, x_n), \ldots, \sigma_{n!} \tau g(x_1, \ldots, x_n)).$$

As σ_i varies over all of S_n, so does $\sigma_i \tau$. Permuting the x_i by τ thus permutes the coordinates in the argument of e; as e is symmetric, it follows that the coefficients of $g^*(x)$ are symmetric in the x_i. Replacing $(x_1, \ldots, x_n)$ by $(\alpha_1, \ldots, \alpha_n)$ thus yields a polynomial in x with coefficients in F, by the corollary to the fundamental theorem of symmetric functions. Although the degree of $g^*(x)$ is large (it is $n!$), it does have one important property: any one of its roots $g(\alpha_1, \ldots, \alpha_n)$ determines all of the others: $\sigma g(\alpha_1, \ldots, \alpha_n) = g(\alpha_{\sigma 1}, \ldots, \alpha_{\sigma n})$ for $\sigma \in S_n$.

In particular, regard $\psi(\omega) = (\alpha_1 + \alpha_2 \omega + \cdots + \alpha_n \omega^{n-1})^n$ as a function of n indeterminates. Then

$$\psi^*(x) = \prod_\sigma (x - \sigma \psi(x_1, \ldots, x_n))$$

is a polynomial with coefficients in $F(x_1, \ldots, x_n)^2$ and replacing $(x_1, \ldots, x_n)$ with $(\alpha_1, \ldots, \alpha_n)$ gives a polynomial in $F[x]$.

One of the roots of $\psi^*(x)$ is $\psi(\omega)$. Assume now that n is prime. If $1 \leq j \leq n-1$, then ω^j is a primitive nth root of unity; hence $\omega^j, \omega^{2j}, \ldots, \omega^{(n-1)j}$ is a permutation, say σ, of $\omega, \omega^2, \ldots, \omega^{n-1}$, and so

$$\psi(\omega^j) = \sigma \psi(\omega).$$

It follows that $\psi(\omega), \psi(\omega^2), \ldots, \psi(\omega^{n-1})$ are roots of $\psi^*(x)$. (This same argument applies to any n if one chooses j relatively prime to n.)

If $g(x_1, \ldots, x_n) \in F[x_1, \ldots, x_n]$, then we can simplify $g^*(x)$ by eliminating repetitions: if $\sigma_i g = \sigma_k g$, throw away one of them.

Definition. A polynomial $g(x_1, \ldots, x_n)$ is r-**valued**[3] where $1 \leq r \leq n!$, if there are exactly r distinct polynomials of the form σg for $\sigma \in S_n$.

Thus, 1-valued functions are symmetric functions; the discriminant is always 2-valued. In the case of a cubic, $\psi(x_1, x_2, x_3) = (x_1 + x_2 \omega + x_3 \omega^2)^3$ is 2-valued and $g(x_1, x_2, x_3) = x_1$ is 3-valued.

----

[2] $F(x_1, \ldots, x_n)$ is the field of fractions of $F[x_1, \ldots, x_n]$; its elements are "rational functions" $a(x_1, \ldots, x_n)/b(x_1, \ldots, x_n)$, where $a, b \in F[x_1, \ldots, x_n]$ and $b \neq 0$.

[3] This is the standard terminology occurring in all the older references. Do not confuse it with modern usage which, for example, calls the relation (not a function) $f(x) = \pm\sqrt{x}$ a 2-valued function.

Plainly, $\psi^*(x)$ should be replaced by its factor $\lambda(x)$ of degree r (where ψ is r-valued) which is obtained from $\psi^*(x)$ by discarding repeated factors. $\lambda(x)$ is called the **Lagrange resolvent** of $f(x)$; this is Lagrange's replacement for the polynomial $\rho(x)$ of degree $n-1$. How can we compute r?

Definition. If $g(x_1, \ldots, x_n) \in F[x_1, \ldots, x_n]$, then

$$G(g) = \{\sigma \in S_n : \sigma g = g\}.$$

Lagrange claimed (but his proof is incomplete) that[4]

$$r = n!/|G(g)|.$$

In particular, an $n!$-valued function $g(x_1, \ldots, x_n)$ is a polynomial with $G(g) = \{1\}$.

There are two ways of regarding a permutation of n letters. The first way is as a word of length n having no repeated letters; the second way is as a bijection. The latter version invites composition: one can multiply two permutations to get a third one. It seems likely that Lagrange was not aware that $G(g)$ is a subgroup of S_n, for he was viewing permutations as words in his proof.

Lagrange did prove a remarkable theorem showing the importance of $G(g)$.

Lagrange Rational Function Theorem. *If $g, h \in F[x_1, \ldots, x_n]$, then $G(h) \subset G(g)$ if and only if g is a rational function of h; that is, there is a rational function $\theta(x) \in F(x)$ with $g = \theta(h)$.*

Corollary 1. *If $g, h \in F[x_1, \ldots, x_n]$, then $G(g) = G(h)$ if and only if each of g and h is a rational function of the other.*

Corollary 2. *If $h \in F[x_1, \ldots, x_n]$ is an $n!$-valued function, then every $g \in F[x_1, \ldots, x_n]$ is a rational function of h.*

Corollary 3. *If $h \in F[x_1, \ldots, x_n]$ is an $n!$-valued function, then each x_i is a rational function of h.*

Corollary 4 (Theorem of Primitive Element). *If $\alpha_1, \ldots, \alpha_n$ are the roots of $f(x) \in F[x]$, then there exists η with $F(\alpha_1, \ldots, \alpha_n) = F(\eta)$.*

---

[4]Here is a modern proof. The group S_n acts on $F[x_1, \ldots, x_n]$ by permuting the variables; $G(g)$ is the stabilizer of g; r is the size of the orbit of g. Theorem A10 gives

$$r = [S_n : G(g)] = n!/|G(g)|.$$

Moreover, there exist rational functions $\theta_i(x) \in F(x)$ with $\alpha_i = \theta_i(\eta)$ for all $i = 1, \ldots, n$.

Proof. Let $h(x_1, \ldots, x_n) \in F[x_1, \ldots, x_n]$ be an $n!$-valued function; for each i, define $g_i(x_1, \ldots, x_n) \in F[x_1, \ldots, x_n]$ by $g_i(x_1, \ldots, x_n) = x_i$. By Corollary 3, there exist rational functions $\theta_i(x) \in F(x)$ with

$$x_i = g(x_1, \ldots, x_n) = \theta_i(h(x_1, \ldots, x_n)).$$

Define $\eta = h(\alpha_1, \ldots, \alpha_n)$.   □

Let us summarize this 1770 work of Lagrange. A polynomial $f(x) \in F[x]$ of degree n determines a polynomial ψ of n variables. This polynomial determines a subgroup $G(\psi)$ of S_n; "symmetrizing" ψ gives a polynomial $\psi^*(x) \in F[x]$ whose roots, when n is prime, suffice to find the roots of $f(x)$. Discarding repeated roots of $\psi^*(x)$ leaves the Lagrange resolvent $\lambda(x) \in F[x]$, a polynomial of degree r, and knowledge of one root of $\lambda(x)$ determines the other roots.

Lagrange hoped that his procedure might solve the general polynomial of degree n. On the other hand, his analysis of the quintic led him to an intractible sextic, with no obvious way to find a root, and this discouraged him.

There was progress in the sixty years from Lagrange to Galois. In 1803, Gauss analyzed roots of unity and cyclotomic polynomials (one consequence is the determination of those regular polygons constructible by ruler and compass). Ruffini (1799) and Abel (1824) essentially proved the insolvability of the general quintic (neither proof is correct in all details, but Abel's proof was accepted and Ruffini's was not). In 1829, Abel proved that certain polynomials $f(x)$ are always solvable by radicals: if $\alpha_1, \ldots, \alpha_n$ are the roots of $f(x)$, if there are rational functions $\theta_i(x)$ with $\alpha_i = \theta_i(\alpha_1)$ for all $i = 1, \ldots, n$, and if

$$\theta_i(\theta_j(\alpha_1)) = \theta_j(\theta_i(\alpha_1))$$

for all i, j (in modern language, the Galois group is abelian; this result is the etymology of the adjective). (See [Tignol; p. 316] for more discussion.)

Although group theory did not exist before Galois, there were some results which today can be seen as group theoretic. Ruffini showed that there are no r-valued functions of 5 variables for $r = 3, 4$, and 8; that is, S_5 has no subgroups of index 3, 4, or 8. Abbati (1803) proved that $|G(g)|$ does, indeed, divide $n!$, so that Lagrange's assertion about the degree r is correct. Thus, Abbati proved "Lagrange's Theorem" (Theorem A3) for subgroups of S_n; the full theorem was probably first proved by Galois. Abbati also proved: if $n \geq 5$, then S_n has no subgroups of index 3 or 4; (Theorem A28) A_n is the only subgroup of S_n having index 2. Cauchy (1815) established the calculus of permutations, e.g., decomposition into disjoint cycles; he proved that, for n prime, S_n has no subgroups of index r with $2 < r \leq n$.

Galois knew that some polynomials are solvable by radicals and some are not; it was reasonable that it depends on the roots. The Lagrange resolvent $\lambda(x)$ is not sensitive to this. Indeed, it seems that Lagrange was seeking a formula for the roots of the *general polynomial* $x^n + b_{n-1} + x^{n-1} + \cdots + b_0$: the roots of any particular polynomial $f(x)$ of degree n would be obtained from the "master formula" by substituting the specific coefficients of $f(x)$. (The classical formulas for polynomials of degree ≤ 4 are of this form.) If $f(x) \in F[x]$ has roots $\alpha_1, \ldots, \alpha_n$, Lagrange first regarded $\alpha_1, \ldots, \alpha_n$ as indeterminates, then he formed $\psi(x_1, \ldots, x_n) = (x_1 + \omega x_2 + \omega^2 x_3 + \cdots + \omega^{n-1} x_n)^n$, symmetrized to obtain

$$\psi^*(x_1, \ldots, x_n) = \prod_{\sigma \in S_n} (x - \sigma \psi(x_1, \ldots, x_n)),$$

defined $\lambda(x)$ to be the factor of ψ^* of degree r (in x) which is the product over all distinct polynomials $\sigma \psi$, and finally specialized $(x_1, \ldots, x_n)$ back to $(\alpha_1, \ldots, \alpha_n)$. But even if $\sigma \psi(x_1, \ldots, x_n)$ and $\tau \psi(x_1, \ldots, x_n)$ are distinct polynomials, it can still happen that $\sigma \psi(\alpha_1, \ldots, \alpha_n) = \tau \psi(\alpha_1, \ldots, \alpha_n)$. As a polynomial over $F(x_1, \ldots, x_n)$, the Lagrange resolvent $\lambda(x) = \lambda(x; x_1, \ldots, x_n)$ has distinct roots; as a polynomial over F, $\lambda(x) = \lambda(x; \alpha_1, \ldots, \alpha_n)$ may have repeated roots. One can discard these extra roots but, unfortunately,

$$\{\sigma \in S_n : (\sigma \psi)(\alpha_1, \ldots, \alpha_n) = \psi(\alpha_1, \ldots, \alpha_n)\}$$

may not be a subgroup of S_n and this prevents the generalization of Lagrange's Rational Function Theorem from being true.

Galois jettisoned $\psi(x_1, \ldots, x_n)$ which, after all, works best when the degree n is prime; he replaced it by an $n!$-valued function $V(x_1, \ldots, x_n)$ with an added property: all $(\sigma V)(\alpha_1, \ldots, \alpha_n)$ are distinct (of course, this forces all the α_i to be distinct; this minor point is easily handled by Exercise 41). Let us call (after Edwards) such a function V a **Galois resolvent**[5] of $f(x)$. Galois proved that such resolvents exist; indeed, there are such of the form $V(x_1, \ldots, x_n) = c_1 x_1 + \cdots + c_n x_n$, for suitable $c_1, \ldots, c_n \in F$. Denote $V(\alpha_1, \ldots, \alpha_n)$ by v_1. Since V is $n!$-valued, there are rational functions $\theta_1(x), \ldots, \theta_n(x)$ in $F(x)$ with $\alpha_i = \theta_i(v_1)$ for all i.

The next step ought to be the symmetrization of V: define

$$V^*(x; x_1, \ldots, x_n) = \prod_{\sigma \in S_n} (x - \sigma V(x_1, \ldots, x_n)),$$

and then choose a factor of V^* by discarding repeated roots. Galois did this indirectly. Let $\gamma(x)$ be the irreducible polynomial of v_1 over F, and let $v_1, \ldots, v_m$ be the roots of $\gamma(x)$.

---

[5] Actually, I would prefer that the polynomial $\gamma(x)$ below be called the Galois resolvent, for it is analogous to $\lambda(x)$ whereas V is analogous to ψ.

Recall Exercise 46: If $p(x), h(x) \in F[x]$ are polynomials having a common root and if $p(x)$ is irreducible, then $p(x)$ divides $h(x)$. It follows at once that $\gamma(x)$ divides $V^*(x) = V^*(x; \alpha_1, \ldots, \alpha_n)$, and so each root v_j of $\gamma(x)$ has the form $\sigma V(\alpha_1, \ldots, \alpha_n)$ for some permutation $\sigma \in S_n$. But Galois wanted a more explicit description of σ. Here is an easy generalization of Exercise 46: Let $p(x) \in F[x]$ be an irreducible polynomial and let $\Phi(x) \in F(x)$ be a rational function; if $\Phi(v) = 0$ for some root v of $p(x)$, then $\Phi(v') = 0$ for every root v' of $p(x)$.

Proposition. *Let $f(x) \in F[x]$ have distinct roots $\alpha_1, \ldots, \alpha_n$, and let $v_1, \ldots, v_m$ be as above; let $\alpha_i = \theta_i(v_1)$, where $\theta_i(x) \in F(x)$ for all i. Then for each $j = 1, \ldots, m$, the function*

$$\sigma_j \colon \alpha_i = \theta_i(v_1) \mapsto \theta_i(v_j), \qquad i = 1, \ldots, n,$$

is a permutation of the roots $\alpha_1, \ldots, \alpha_n$.

Proof. Define $\Phi(x) \in F(x)$ by $\Phi(x) = f(\theta_i(x))$. Now $\Phi(v_1) = f(\theta_i(v_1)) = f(\alpha_i) = 0$; since $\gamma(x)$ is irreducible, the generalized Exercise 46 shows that $0 = \Phi(v_j) = f(\theta_i(v_j))$; that is, $\theta_i(v_j)$ is a root of $f(x)$, hence is one of the α's. To see that σ_j is a permutation, it suffices to prove it is an injection. Suppose that $\theta_i(v_j) = \theta_k(v_j)$. Now $\Phi(x) = \theta_i(x) - \theta_k(x)$ is a rational function with $\Phi(v_j) = 0$; it follows that $0 = \Phi(v_1) = \theta_i(v_1) - \theta_k(v_1) = \alpha_i - \alpha_k$. Since all the roots of $f(x)$ are distinct, $i = k$, as desired. $\quad\square$

Galois defined the Galois group of $f(x)$ as

$$\mathrm{Gal}(f) = \{\text{all } \sigma_j \colon \alpha_i = \theta_i(v_1) \mapsto \theta_i(v_j)\}.$$

This is the beginning of Galois's 1831 paper in which he characterizes polynomials solvable by radicals as those having solvable Galois group. (For a proof that this definition is equivalent to the modern one in terms of automorphisms, see [Tignol, p. 329].)

Subtle group theoretic clues were in the air, but only Galois recognized their significance; developing them, he invented group theory and solved the mystery of the roots of polynomials. This is even more impressive when we realize that this is no less than the birth of modern algebra.

References

Artin, E., *Galois Theory* (second edition), Notre Dame, 1955.

Burnside, W.S., and Panton, A.W., *The Theory of Equations*, vol. II, Longmans, Green, 1899.

Chase, S., Harrison, D., and Rosenberg, A., Galois Theory and Cohomology of Commutative Rings, *Mem. Amer. Math. Soc.*, 1965.

Dehn, E., *Algebraic Equations*, Columbia University Press, 1930.

Edwards, H.M., *Galois Theory*, Springer, 1984.

Gaal, L., *Classical Galois Theory with Examples* (fourth edition), Chelsea, 1988.

Hadlock, C.R., Field Theory and Its Classical Problems, *Math. Assn. Amer.*, 1978.

Jacobson, N., Structure of Rings, *Amer. Math. Soc.*, 1956.

Jacobson, N., *Basic Algebra I*, Freeman, 1974.

Kaplansky, I., *An Introduction to Differential Algebra*, Hermann, 1957.

Kaplansky, I., *Fields and Rings* (second edition), University Chicago Press, 1974.

Miller, G.A., Blichfeldt, H.F., and Dickson, L.E., *Theory and Applications of Finite Groups*, Wiley, 1916 (Dover, 1961).

Netto, E., *Theory of Substitutions*, 1882, reprinted Chelsea, 1961.

Tignol, J.-P., *Galois's Theory of Algebraic Equations*, Wiley, 1988.

van der Waerden, B.L., *Modern Algebra I*, Ungar, 1953.

van der Waerden, B.L., *A History of Algebra*, Springer, 1985.

Index

See Appendix 1 for definitions of group theoretic terms.